Lecture Notes in Artificial Intelligence　10571

Subseries of Lecture Notes in Computer Science

More information about this series at http://www.springer.com/series/1244

Vicenç Torra · Yasuo Narukawa
Aoi Honda · Sozo Inoue (Eds.)

Modeling Decisions for Artificial Intelligence

14th International Conference, MDAI 2017
Kitakyushu, Japan, October 18–20, 2017
Proceedings

Springer

Editors
Vicenç Torra
University of Skövde
Skövde
Sweden

Yasuo Narukawa
Toho Gakuen
Kunitachi, Tokyo
Japan

Aoi Honda
Kyushu Institute of Technology
Iizuka, Fukuoka
Japan

Sozo Inoue
Kyushu Institute of Technology
Kitakyushu-shi, Fukuoka
Japan

ISSN 0302-9743 ISSN 1611-3349 (electronic)
Lecture Notes in Artificial Intelligence
ISBN 978-3-319-67421-6 ISBN 978-3-319-67422-3 (eBook)
DOI 10.1007/978-3-319-67422-3

Library of Congress Control Number: 2017952883

LNCS Sublibrary: SL7 – Artificial Intelligence

Printed on acid-free paper

This Springer imprint is published by Springer Nature
The registered company is Springer International Publishing AG
The registered company address is: Gewerbestrasse 11, 6330 Cham, Switzerland

Preface

This volume contains papers presented at the 14th International Conference on Modeling Decisions for Artificial Intelligence (MDAI 2017), held in Kitakyushu, Japan, October 18–20, 2017. This conference followed MDAI 2004 (Barcelona, Spain), MDAI 2005 (Tsukuba, Japan), MDAI 2006 (Tarragona, Spain), MDAI 2007 (Kitakyushu, Japan), MDAI 2008 (Sabadell, Spain), MDAI 2009 (Awaji Island, Japan), MDAI 2010 (Perpignan, France), MDAI 2011 (Changsha, China), MDAI 2012 (Girona, Spain), MDAI 2013 (Barcelona, Spain), MDAI 2014 (Tokyo, Japan), MDAI 2015 (Skövde, Sweden), and MDAI 2016 (Sant Julià de Lòria, Andorra) with proceedings also published in the LNAI series (Vols. 3131, 3558, 3885, 4617, 5285, 5861, 6408, 6820, 7647, 8234, 8825, 9321, and 9880).

The aim of this conference was to provide a forum for researchers to discuss theory and tools for modeling decisions, as well as applications that encompass decision making processes and information fusion techniques.

The organizers received 30 papers from 12 different countries, 18 of which are published in this volume. Each submission received at least two reviews from the Program Committee and a few external reviewers. We would like to express our gratitude to them for their work. This volume also includes some of the plenary talks.

The conference was supported by the Kyushu Institute of Technology, the Kitakyushu Convention & Visitors Association, the Japan Society for Industrial and Applied Mathematics, the Operations Research Society of Japan, the Japan Society for Fuzzy Theory and Intelligent Informatics (SOFT), the European Society for Fuzzy Logic and Technology (EUSFLAT), the Catalan Association for Artificial Intelligence (ACIA), and the UNESCO Chair in Data Privacy.

July 2017

Vicenç Torra
Yasuo Narukawa
Aoi Honda
Sozo Inoue

Organization

General Chairs

Aoi Honda   Kyushu Institute of Technology, Kitakyushu, Japan
Sozo Inoue   Kyushu Institute of Technology, Kitakyushu, Japan

Program Chairs

Vicenç Torra   University of Skövde, Skövde, Sweden
Yasuo Narukawa   Toho Gakuen, Tokyo, Japan

Advisory Board

Didier Dubois   Institut de Recherche en Informatique de Toulouse, CNRS, France
Lluis Godo   IIIA-CSIC, Catalonia, Spain
Kaoru Hirota   Beijing Institute of Technology; JSPS Beijing Office, China
Janusz Kacprzyk   Systems Research Institute, Polish Academy of Sciences, Poland
Sadaaki Miyamoto   University of Tsukuba, Japan
Yoshiaki Okazaki   Kyushu Institute of Technology, Kitakyushu, Japan
Michio Sugeno   Tokyo Institute of Technology, Japan
Ronald R. Yager   Machine Intelligence Institute, Iona College, NY, USA

Program Committee

Eva Armengol   IIIA-CSIC, Catalonia, Spain
Edurne Barrenechea   Universidad Pública de Navarra, Spain
Gloria Bordogna   Consiglio Nazionale delle Ricerche, Italy
Humberto Bustince   Universidad Pública de Navarra, Spain
Francisco Chiclana   De Montfort University, UK
Susana Díaz   Universidad de Oviedo, Spain
Josep Domingo-Ferrer   Universitat Rovira i Virgili, Catalonia, Spain
Jozo Dujmovic   San Francisco State University, California
Yasunori Endo   University of Tsukuba, Japan
Zoe Falomir   Universität Bremen, Germany
Katsushige Fujimoto   Fukushima University, Japan
Michel Grabisch   Université Paris I Panthéon-Sorbonne, France
Enrique Herrera-Viedma   Universidad de Granada, Spain
Aoi Honda   Kyushu Institute of Technology, Japan
Masahiro Inuiguchi   Osaka University, Japan

Simon James	Deakin University, Australia
Yuchi Kanzawa	Shibaura Institute of Technology, Japan
Petr Krajča	Palacky University Olomouc, Czech Republic
Marie-Jeanne Lesot	Université Pierre et Marie Curie (Paris VI), France
Xinwang Liu	Southeast University, China
Jun Long	National University of Defense Technology, China
Jean-Luc Marichal	University of Luxembourg, Luxembourg
Radko Mesiar	Slovak University of Technology, Slovakia
Andrea Mesiarová-Zemánková	Slovak Academy of Sciences, Slovakia
Tetsuya Murai	Hokkaido University, Japan
Toshiaki Murofushi	Tokyo Institute of Technology, Japan
Guillermo Navarro-Arribas	Universitat Autònoma de Barcelona, Catalonia, Spain
Gabriella Pasi	Università di Milano-Bicocca, Italy
Sandra Sandri	Instituto Nacional de Pesquisas Espaciais, Brazil
László Szilágyi	Sapientia-Hungarian Science University of Transylvania, Hungary
Aida Valls	Universitat Rovira i Virgili, Catalonia, Spain
Zeshui Xu	Southeast University, China
Yuji Yoshida	University of Kitakyushu, Japan

Local Organizing Committee Chair

Hiroshi Sakai	Kyushu Institute of Technology, Kitakyushu, Japan

Local Organizing Committee

Hideaki Kawano	Kyushu Institute of Technology, Japan
Makoto Ohki	National Institute of Technology, Kumamoto College, Japan

Additional Referees

Luis del Vasto
Sara Ricci
Sergio Martnez
Juan A. Rodriguez-Aguilar
Jordi Casas

Supporting Institutions

Kyushu Institute of Technology, Japan
The Kitakyushu Convention & Visitors Association, Japan
The Japan Society for Industrial and Applied Mathematics
The Operations Research Society of Japan
The Japan Society for Fuzzy Theory and Intelligent Informatics (SOFT)
The European Society for Fuzzy Logic and Technology (EUSFLAT)
The Catalan Association for Artificial Intelligence (ACIA)
The UNESCO Chair in Data Privacy

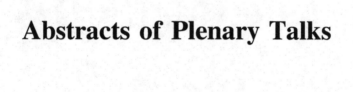

Abstracts of Plenary Talks

Integration of Non-Additive Probabilities: Aggregation When Information Is Incomplete

Ehud Lehrer

Department of Statistics and Operations Research, Tel-Aviv University, Israel

Abstract. Quite often decision makers have only partial information about the underlying uncertainty. This might happen, for instance, when information about the subject matter is obtained from different surveys/resources. We model such an information as a non-additive probability. Consider a decision maker who has to choose between two portfolios, or between two groups of engineers, based on incomplete information about the uncertainty of the market, or about the productivity of the groups. How would the decision maker evaluate the expected return from each portfolio or expected productivity from each group? We present different schemes of aggregation with respect to non-additive probabilities. These schemes might serve as decision tools in many fields, such as financial markets, production and more.

Sparsity Methods for Estimation and Control

Masaaki Nagahara

Institute of Environmental Science and Technology,
The University of Kitakyushu, Hibikino, Wakamatsu-ku, Kitakyushu,
Fukuoka 808-0135, Japan

Abstract. Recently, sparsity has been playing a central role in signal processing, machine learning, and data science. Here we consider a problem of reconstructing (or learning) a signal (or a function) from observed data, which may be under-sampled and disturbed by noise. To address this problem, a method called sparse modeling, also known as compressed sensing, has become a hot topic. In this talk, I will give a brief introduction to sparse modeling for signal estimation, and its applications to control. In particular, I will give an introduction to "maximum hands-off control," which has the minimum support length among all feasible solutions for saving energy and reducing CO_2 emissions in control systems.

Stream Data Compression and Its Applications

Hiroshi Sakamoto

Kyushu Institute of Technology, Japan

Abstract. Social networking service and sensing device have become more and more popular in recent years and data flow never stop to increase. Examples are genome sequences of same species, version controlled documents, and source codes in repositories. Since such a data is usually highly-compressible, adopting data compression techniques is a suitable way to process it. In addition, in order to catch up the speed of data grow, there is a strong demand for stream data compression, that is, fully online and really scalable compression. In this talk, I would like to focus on lossless data compression and introduce several state-of-the-art technologies for stream data compression including their applications.

Provenance and Privacy

Vicenç Torra[1], Guillermo Navarro-Arribas[2], David Sanchez-Charles[3],
and Victor Muntés-Mulero[3]

[1] School of Informatics, University of Skövde, Skövde, Sweden
vtorra@his.se
[2] Department of Information and Communication Engineering,
Universitat Autònoma de Barcelona, Spain
guillermo.navarro@uab.cat
[3] CA Technologies, Barcelona, Spain
{david.sanchez,victor.muntes}@ca.com

Abstract. This paper presents an overview of current needs on data provenance
and data privacy, and discusses state-of-the-art results in this area. The paper
highlights the difficulties that we need to face and finishes with some lines that
require further work.

Contents

Data Privacy and Security

Data Mining and Applications

Invited Paper

Provenance and Privacy

Vicenç Torra[1], Guillermo Navarro-Arribas[2(✉)], David Sanchez-Charles[3],
and Victor Muntés-Mulero[3]

[1] School of Informatics, University of Skövde, Skövde, Sweden
vtorra@his.se
[2] Department of Information and Communication Engineering,
Universitat Autònoma de Barcelona, Barcelona, Spain
guillermo.navarro@uab.cat
[3] CA Technologies, Barcelona, Spain
{david.sanchez,victor.muntes}@ca.com

Abstract. This paper presents an overview of current needs on data provenance and data privacy, and discusses state-of-the-art results in this area. The paper highlights the difficulties that we need to face and finishes with some lines that require further work.

1 Introduction

Data provenance is the technology that permits to have the history of the data till present, where data comes from and which processes were applied to the data. There are multiple reasons why this information is relevant. Applications of data provenance include scientific data and e-science, archival, accounting (financial data), medical data, and pharmaceutical provenance. Data provenance permits us to trace the data from their origins, calibrate systems, and reproduce experiments. Companies hold large amounts of information about customers that have been gathered with the purpose of providing personalized services. Other data is recorded for a future and undefined use. To hold and process this data is important to increase the competitive advantage of a company.

When data is held, data privacy regulations need to be taken into account. Data access has to be implemented so that sensitive data is not disclosed to unauthorized people. Data privacy methods and access control protocols are being developed for this purpose [26]. Two important aspects have irrupted recently within data privacy. On the one hand, European legislation put into force the right to be forgotten. That is, individuals can require companies to delete the data they hold that relate to them. Recall the general principle established by the European Court of Justice in May 2014 in the Google Spain v. AEPD and Mario Costeja Gonzalez process granting individuals the right to delete records concerning themselves from Google's database. After this ruling, Google has processed 421,949 requests to remove almost 1.5 million links [9]. Requests from Sweden have caused the removal of 41,618 links. On the other hand, individuals have also the right to amend their own information. The new EU General Data Protection Regulation that was adopted on 27 April 2016 and it shall apply

© Springer International Publishing AG 2017
V. Torra et al. (Eds.): MDAI 2017, LNAI 10571, pp. 3–11, 2017.
DOI: 10.1007/978-3-319-67422-3_1

from 25 May 2018 consolidates these two rights and it includes as one of the key changes the right to be forgotten and an easier access to personal information.[1] It is important to note that the deletion or modification of the data supplied by individuals is often not enough: data is typically aggregated and used to build models and make decisions. Deletion and amendment may require the reconsideration of inferences made from these data elements, and reconsideration of the knowledge extracted from the data. Industry needs to get ready and have software control data provenance to know who created and modified data. This software has to be able to erase or modify data when necessary. Software also needs to help the industry to know who contributed to data models and decisions, and help to update models and decisions when they are affected by deletions and updates. Only in this way, data rich business can run smoothly under the implementation of the above mentioned individual rights.

Data provenance, also known as data and information lineage, provides tools to know where data comes from (i.e., sources that have contributed) and how these data have been processed. Provenance structures, which in short are annotations on the data, can be rather complex as data elements can be obtained as the result of the integration of several information sources, and/or the application of complex models which in turn have been obtained from other data. The advantages of data provenance is that it improves data quality, permits accountability, and it is essential to implement the right to be forgotten and the right to amend. Nevertheless, the implementation of data provenance poses some problems as they are complex structures that may duplicate (or more) the size of a database when implemented at value level. Provenance structures can contain sensitive information that may require the implementation of appropriate data privacy and access control policies. In addition, integrity should be guaranteed to provenance structures to avoid their arbitrary modification. The need for integrity is specially relevant in distributed systems, when data flows from one business to another and we need to ensure data provenance integrity. Secure data provenance focus on this issue. Data provenance, which is neither standard nor fully implemented in regular-size databases, is still more complex in the context of big data. This is due to the amount of information to be dealt with and the large number of inferences extracted from a single data element. Provenance annotations, and access control methods increase the computational cost with respect to space and execution time. The need for ensuring data provenance in big data has been underlined in a few reports and research papers. See e.g., the USA Report to the President [22].

In addition, in the era of big data and online social networks, data provenance is useful to help users to assess the validity and trust of the information. For instance, it can help to identify rumormongers and disinformation centers. Several important cases have been identified recently were rumors and fake information were produced to intoxicate the political discussion and provoke international political conflicts. Some of these news generated in social networks even

---

[1] Regulation (EU) 2016/679 of the European Parliament and of the Council: http://eur-lex.europa.eu/legal-content/EN/TXT/?uri=CELEX:32016R0679.

reached the press. Recall the two cases concerning to Sweden politicians and the war in Ukraine. First, the fake letter attributed to the Minister of Defence Peter Hultqvist in February 2015 [25] and the letter attributed to Tora Holst (chief prosecutor at the International Public Prosecution Office in Stockholm) in September 2015 which reached CNN iReport [1].

In this paper, we present a review of techniques, methods and characteristics of data provenance, we focus on the aspects related to big data. In Sect. 2 we discuss the representation of provenance. In Sect. 3 the characteristics of secure data provenance. Section 4 is a review of systems for both data provenance and secure data provenance. Section 5 is about standards related to data provenance. In Sect. 6 we open two research lines related to the usage of data provenance and privacy in combination with machine learning techniques. The paper finishes with some conclusions.

2 Provenance Representation

Data provenance can be seen as metadata or as an annotation of the data. That is, data is expanded with information of the processes that has led to this data. Provenance can be coarse-grained or fine-grained. That is, we can have information on how a bunch of data (i.e., files or databases have been produced) or we can have information particularized at the record or even at the value level. Fine-grained provenance is what makes provenance useful, as it is only in this case that we have detailed information on how any data element has been produced. E.g., we can know who entered the temperature (fever) of a patient, or in which store our client claimed for a discount. There are different ways to represent data provenance. There are two types of provenance. They are *where provenance* and *why provenance*. *Where provenance* describes the origin of the data, and *why provenance* the process that generated the data. A data element in a database typically proceeds from the combination of previous data elements by means of certain processing functions. Therefore, we need a structure to represent the transformations. The most common approaches are chains and graphs. In chains we assume that the processes applied to the data are sequential and that there is a single origin. In contrast, in graphs we consider that data has multiple origins and they are the result of the application of functions that have inferred new values from multiple data elements. Tree-like structures also permit us to represent combination of data from multiple sources but is unable to represent a combination of data from the same source after the application of different partial procedures. The representation by means of graphs is naturally more flexible than using sequences or trees. [4,29] are probably the first papers to use graphs for this purpose. [11,12,16] use chains. As provenance graphs represent a causality [4] relationship and nodes are actions in a given time, they do not include cycles. I.e., they are directed acyclic graphs. Other provenance structures may be conceivable.

2.1   Nodes and Operations on Data Provenance

Following [29], we consider graphs in which a compact notation is used to represent all types of processes (*why* and *where provenance*). In short, nodes represent the application of a process to generate a new data element. Each node is a tuple with the following structure:

$$(seqID, p, \{(A_1, v_1), \ldots, (A_n, v_n)\}, (A, v))$$

where $seqID$ is an identifier and can include a time-stamp (in any case, it permits to order nodes with respect to time), p is the subject who applied a process, $\{(A_1, v_1), \ldots, (A_n, v_n)\}$ is the set of input in which (A_i, v_i) is a pair consisting on an object A_i and its value v_i, and (A, v) represents the output object and its value.

In this notation *where provenance* is represented by means of an empty set instead of $\{(A_1, v_1), \ldots, (A_n, v_n)\}$. This notation is preferable with respect to others in which only the operation is represented. Explicit representation of input and output information can be useful for black-box operations and non-deterministic functions.

Computation of new values requires the expansion of the provenance graph, adding new nodes. As previous operations cannot be revisited the graph is an immutable structure.

2.2   Properties of Data Provenance

We can distinguish two main properties that a data provenance system needs to satisfy:

- **Completeness.** That is, that all actions that are relevant to computation should be detected and represented in the provenance structure. Note that this is not always easy, because some operations as e.g. cut & paste or manual copy can exclude relevant provenance information.
- **Efficiency.** Data provenance introduces an overhead to the data. Fine-grained provenance can double (or more) the size of a database. In addition, operations on the provenance structure need to be efficient because they also introduce an overhead on the computation time.

Note that although both issues need to be implemented, efficiency is a much more relevant and complex issue in the framework of secure data provenance for big data. Efficient methods have to be designed to satisfy the four properties discussed here.

3   Secure Data Provenance

Secure provenance was introduced to ensure security and privacy to provenance data. Observe that provenance data is sensitive. It may contain information on who and when data was updated. E.g., knowing that a certain doctor has modified data from a patient can lead to disclosure on who is the doctor of whom, what type of illness the patient has, and at what time the patient was at the hospital. Files and databases typically flow within departments and between companies. It is specially important to ensure that these third parties cannot access confidential information contained in the data provenance, whilst allowing them to work with the factual data and update the provenance structure itself. For example, this would allow to perform analysis on the medical data, preserving patient privacy. Hence, provenance data needs to follow these databases and this has to be done ensuring e.g., provenance integrity. Secure data provenance focuses on these type of problems. A few properties have been established as a requirement for secure data provenance [11,12,29]:

- **Distributed.** When databases flow through untrusted environments, and provenance data is associated to them, we need secure data provenance systems to be defined so that they work in a distributed environment. We cannot use a centralized approach with trusted hardware.
- **Integrity.** In distributed environments it is important that nobody can forge provenance data. Provenance data is transmitted and provenance structures are modified to add the new processes applied to the data. Nevertheless, as stated above the structure is immutable and no adversary can be granted to change any part of it. In addition, the provenance system should not allow the modification of a value without expanding the provenance structure. Finally, deletion of provenance data should not cause that a record of the database is unreadable. Additional aspects to be taken into account is to consider collusions of intruders (that coalitions of intruders should not be able to attack integrity), repudiation (that intruders should not be able to repudiate a record as it was not theirs) or creating forged structures (intruders should not be able to create new provenance structures).
- **Availability.** We are interested in providing security mechanisms to ensure provenance data availability. Auditors should be able to access provenance information in a secure, fast and reliable manner to perform any required operation, e.g. verify the integrity of an ownership sequence without knowing the individual records.
- **Privacy and confidentiality.** We need to ensure that disclosure does not take place, and this is needed for both the database and the provenance data. Only authorized users can access the information.

These properties need to be combined with the two properties discussed in the previous section that are general for any provenance system. They are, completeness and efficiency.

All these properties are relevant in the context of big data provenance. Big data is often distributed as different information sources can contribute in a computation or in a decision. Therefore, integrity is a basic aspect. We need that provenance structures are not modified at will, and we need to be sure that only permitted operations are applied to them. Availability is then not only a requirement for auditors but also for the subjects from which the data has been extracted. In order that individuals can access and apply the right to delete or amend a record, they need to be able to know where is their data or if a certain record contains data that has been generated from their own data.

4 Data Provenance and Secure Data Provenance Systems

Although data provenance support can be found in some relational database and data management products [3,19], the development of secure data provenance is very scarce. Moreover, current approaches to support provenance in big data management and storage systems (Hadoop, Spark, HBase, etc.) are in a very early stage, usually limited to academic projects that solve very specific problems.

RAMP [20] provides a wrapper approach for Hadoop by wrapping map and reduce functions without requiring the modification of the Hadoop core. Similarly, HadoopProv [2] also captures provenance information in map and reduce functions but attempts to minimize its overhead by deferring the construction of the provenance graph in a later stage. Although the later approach introduces a lower overhead in runtime, the first one will respond faster to provenance queries, and thus to provenance related access control. In the same line Titian [14] is an extension to Spark adding interactive data provenance to data transformations. Some other systems also address some limited provenance support in Hadoop-like systems towards providing debugging information. For example, Newt [15] supports *why provenance* for debugging purposes.

In general there is a clear lack of sound provenance support in big data processing systems. Not only their functionality is very limited, but there is no support for secure and privacy-preserving provenance.

5 Standards and Data Provenance

The W3C produced the PROV specifications [8] which define a model, serialization and related tools for the interchange of provenance information, specifically focused (but not limited) to the Web. These specifications have not yet been adopted by provenance aware databases [18], but PROV has been widely used as a standard for the exchange of provenance information. PROV provides a standardized model based on relations among agents, entities and activities, which can be expressed in several formats. The specifications cover a wide range of different aspects from ontologies (OWL2, RDF, ...), human readable notations, constraints mechanisms for accessing and querying provenance, etc.

PROV comes from the semantic web community, which departs and attempts to cover work from previous efforts from similar proposals. This includes Dublin Core, the Open Provenance Model [17], Provernir Ontology [23], Provenance Vocabulary [10], SWAN (Semantic Web Applications in Neuromedicine) Ontology [5], among others.

In the eHealth field an initiative from the S&I Framework [27] is currently working towards a standardization of data provenance in the context of Health IT. HL7 [13] will incorporate this provenance support for CDA

(Clinical Document Architecture) documents (HL7), which includes typical clinical documents and data exchanges. This work is still in a very early stage, especially regarding the security and privacy surrounding the provenance data, which in the case of Health IT is of key relevance. Although they are not specifically addressing big data, the development of new approaches to deal with secure provenance in big data will surely contribute to better understand the problem and contribute to successfully extend initiative such as this one.

6   Data Provenance and Privacy on Multi-actor Systems

In most cases, data owners cannot couple with most of the processes needed in order to get the most benefits from the data, and the services of third parties are usually needed. Hence, files and databases are shared between companies, difficulting compliance of data regulations, in particular related to the right to be forgotten and data amendment, and posing a potential violation of user privacy. It is particularly interesting in the machine learning field, in which reducing the number of model recalculations is desirable.

Data sharing or publication must ensure privacy of the original owners. There are three main privacy models in data privacy: k-anonymity [24], differential privacy [7] and reidentification [6, 28]. Reidentification and k-anonymity focus on avoiding identity disclosure. That is, the goal of the intruder is to link a record of a file with an individual (or company). In k-anonymity identity disclosure is avoided having for any information of the intruder, k records that match this information. Differential privacy has a different perspective. It is defined in terms of a query. Then, the goal of differential privacy is that the output of a query does not change significantly when a record is added/removed from a database. As differential privacy focus on a given query, it can be seen as focusing on attribute disclosure (i.e., the goal of the intruder is to increase their knowledge on the value of an attribute for a given record).

In the context of data provenance, differential privacy [7] is interesting as it studies algorithms that try to return similar results when applied to two databases that only differ on a single record. Hence, when individuals require the deletion of a record, if data inferences were made using a differentially-private algorithm we assess if such inference is somehow still valid. Differential privacy usually proceeds by means of a randomization of the output. This means that different executions of a differentially private algorithm on the same data typically results into different outputs. This may be a problem in real practice because when building a model we need to know that the model produced is near to the optimal one. Typically, differential privacy does not allow to repeat the query to avoid the intruder having an accurate estimate of the distribution of the output, so it may be difficult to evaluate the accuracy of the obtained calculation. In addition, previous work in differential privacy does not consider sequences of deletions and amendments.

Provenance structures are not the only data annotations. Sticky policies [21], a concept arisen in the context of data privacy, is another type of data annotation that focuses on the conditions and constraints that describe how data should be treated so that it is compliant with and enforces current data privacy requirements. An important novel research area may be defining such conditions and constraints based on the provenance of the data itself, by only allowing certain groups of individuals, for instance, to access data after a certain level of anonymization or aggregation is guaranteed.

7   Conclusions

In this paper, we have presented an overview of the main aspects related to data provenance and data privacy. This is an area of increasing interests due to current needs and regulations. We have reviewed in Sect. 6 some lines of research related to both areas.

Acknowledgements. Partially supported by Vetenskapsrådet project: "Disclosure risk and transparency in big data privacy" (VR 2016-03346); and Spanish and Catalan governments with projects TIN2014-55243-P and 2014SGR-691 respectively.

References

1. Ahlborg, K.: Vem har skrivit det falska brevet om Sverige? Aftonbadet 12th September 2015. http://www.aftonbladet.se/nyheter/article21407033.ab
2. Akoush, S., Sohan, R., Hopper, A.: HadoopProv: towards Provenance as a first class citizen in MapReduce. In: Presented as Part of the 5th USENIX Workshop on the Theory and Practice of Provenance (2013)
3. Arab, B., Gawlick, D., Radhakrishnan, V., Guo, H., Glavic, B.: A generic provenance middleware for queries, updates, and transactions. In: 6th USENIX Workshop on the Theory and Practice of Provenance (2014)
4. Braun, U., Shinnar, A., Seltzer, M.: Securing provenance. In: Proceedings of the HOTSEC (2008)
5. Ciccarese, P., Wu, E., Wong, G., Ocana, M., Kinoshita, J., Ruttenberg, A., Clark, T.: The SWAN biomedical discourse ontology. J. Biomed. Inform., Semant. Mashup Biomed. Data **41**, 739–751 (2008). doi:10.1016/j.jbi.2008.04.010
6. Domingo-Ferrer, J., Torra, V.: A quantitative comparison of disclosure control methods for microdata. In: Doyle, P., Lane, J.I., Theeuwes, J.J.M., Zayatz, L. (eds.) Confidentiality, Disclosure and Data Access: Theory and Practical Applications for Statistical Agencies, North-Holland, pp. 111–134 (2001)
7. Dwork, C.: Differential privacy. In: Bugliesi, M., Preneel, B., Sassone, V., Wegener, I. (eds.) ICALP 2006. LNCS, vol. 4052, pp. 1–12. Springer, Heidelberg (2006). doi:10.1007/11787006_1
8. Gil, Y., Miles, S. (eds.): PROV Model Primer (2013)
9. Google.: European privacy requests for search removals (2016). https://www.google.com/transparencyreport/removals/europeprivacy/?hl=en
10. Hartig, O., Zhao, J.: Publishing and consuming provenance metadata on the web of linked data. In: McGuinness, D.L., Michaelis, J.R., Moreau, L. (eds.) IPAW 2010. LNCS, vol. 6378, pp. 78–90. Springer, Heidelberg (2010). doi:10.1007/978-3-642-17819-1_10
11. Hasan, R., Sion, R., Winslett, M.: Introducing secure provenance: problems and challenges. In: Proceedings of the StorageSST. ACM (2007)

12. Hasan, R., Sion, R., Winslett, M.: The case of the fake Picasso: preventing history forgery with secure provenance. In: Proceedings of the FAST (2009)
13. HL7 Standards Product Brief - HL7 CDA? R2 Implementation Guide: Data Provenance, Release 1 - US Realm [WWW Document], n.d. http://www.hl7.org/implement/standards/product_brief.cfm?product_id=420. Accessed 04 Apr 2016
14. Interlandi, M., Shah, K., Tetali, S.D., Gulzar, M.A., Yoo, S., Kim, M., Millstein, T., Condie, T.: Titian: data provenance support in Spark. Proc. VLDB Endow 9, 216–227 (2015)
15. Logothetis, D., De, S., Yocum, K.: Scalable Lineage Capture for Debugging DISC Analytics, pp. 1–15. ACM Press, New York (2013). doi:10.1145/2523616.2523619
16. McDaniel, P., Butler, K., Sion, R., Zadok, E., Winslett, M.: Towards a secure and efficient system for end-to-end provenance. In: Proceedings of the TAPP (2010)
17. Moreau, L., Freire, J., Futrelle, J., McGrath, R.E., Myers, J., Paulson, P.: The open provenance model: an overview. In: Freire, J., Koop, D., Moreau, L. (eds.) IPAW 2008. LNCS, vol. 5272, pp. 323–326. Springer, Heidelberg (2008). doi:10.1007/978-3-540-89965-5_31
18. Niu, X., Glavic, B., Gawlick, D., Liu, Z.H., Krishnaswamy, V., Radhakrishnan, V.: Interoperability for Provenance-aware Databases using PROV and JSON. In: 7th USENIX Workshop on the Theory and Practice of Provenance (2015)
19. Oracle Enterprise Metadata Management Overview [WWW Document], n.d. http://www.oracle.com/technetwork/middleware/oemm/overview/index.html. Accessed 04 Apr 2016
20. Park, H., Ikeda, R., Widom, J.: RAMP: a system for capturing and tracing provenance in MapReduce workflows. In: Presented at the 37th International Conference on Very Large Data Bases, Stanford InfoLab, Seattle, Washington (2011)
21. Pearson, S., Mont, M.C.: Sticky Policies: An approach for managing privacy across multiple parties, Computer, pp. 60–68 (2011)
22. PCAST.: Big data and privacy: a technological perspective, President's Council of Advisors on Science and Technology, Executive office of the president of the United States (2014)
23. Sahoo, S.S., Sheth, A.P.: Provenir ontology: towards a framework for eScience provenance management. Microsoft eScience Workshop (2009)
24. Samarati, P.: Protecting Respondents' Identities in Microdata Release. IEEE Trans. Knowl. Data Eng. 13(6), pp. 1010–1027 (2001)
25. Thurfjell, K.: Sverige erbjuder sig sälja vapen till Ukraina i förfalskat brev, Svenska Dagbladet (2015). http://www.svd.se/sverige-erbjuder-sig-salja-vapen-till-ukraina-i-forfalskat-brev
26. Torra, V., Navarro-Arribas, G.: Data Privacy. WIREs Data Min. Knowl. Discov. 4(4), pp. 269–280 (2014)
27. http://wiki.siframework.org/Data+Provenance+Initiative
28. Winkler, W.E.: Re-identification methods for masked microdata. In: Domingo-Ferrer, J., Torra, V. (eds.) PSD 2004. LNCS, vol. 3050, pp. 216–230. Springer, Heidelberg (2004). doi:10.1007/978-3-540-25955-8_17
29. Zhang, J., Chapman, A., LeFevre, K.: Do you know where your data's been? – tamper-evident database provenance. In: Jonker, W., Petković, M. (eds.) SDM 2009. LNCS, vol. 5776, pp. 17–32. Springer, Heidelberg (2009). doi:10.1007/978-3-642-04219-5_2

Aggregation Operators, Fuzzy Measures and Integrals

Comparison of Risk Averse Utility Functions on Two-Dimensional Regions

Yuji Yoshida[✉]

Faculty of Economics and Business Administration, University of Kitakyushu,
4-2-1 Kitagata, Kokuraminami, Kitakyushu 802-8577, Japan
yoshida@kitakyu-u.ac.jp

Abstract. Weighted quasi-arithmetic means on two-dimensional regions are demonstrated, and risk averse conditions are discussed by the corresponding utility functions. For two utility functions on two-dimensional regions, we introduce a concept that decision making with one utility is more risk averse than decision making with the other utility. A necessary condition and a sufficient condition for the concept are demonstrated by their utility functions. Several examples are given to explain them.

1 Introduction

Weighted quasi-arithmetic means are important concept for mathematical theory such as the mean value theorems, and it is a fundamental tool for subjective estimation regarding information in management science, artificial intelligence and so on. Weighted quasi-arithmetic means of an interval are given mathematically by aggregation operations (Kolmogorov [4], Nagumo [6] and Aczél [1]). Bustince et al. [2] discussed aggregation operations on two-dimensional OWA operators, and Labreuche and Grabisch [5] demonstrated Choquet integral for aggregation in multicriteria decision making, and Torra and Godo [7] studied continuous WOWA operators for defuzzification. In micro-economics, subjective estimations with preference relations are formulated as utility functions (Fishburn [3]). From the view point of utility functions, Yoshida [8,9] have studied the relations between weighted quasi-arithmetic means on an interval and decision maker's behavior regarding risks. In one-dimensional cases, for twice continuously differentiable strictly increasing functions $\varphi, \psi : [a, b] \mapsto \mathbb{R}$ as decision makers' *utility functions* and a continuous function $\omega : [a, b] \mapsto (0, \infty)$ as a *weighting function*, *weighted quasi-arithmetic means* μ and ν on a closed interval $[a, b]$ are real numbers satisfying

$$\varphi(\mu) \int_a^b \omega(x)\,dx = \int_a^b \varphi(x)\,\omega(x)\,dx, \tag{1.1}$$

$$\psi(\nu) \int_a^b \omega(x)\,dx = \int_a^b \psi(x)\,\omega(x)\,dx \tag{1.2}$$

© Springer International Publishing AG 2017
V. Torra et al. (Eds.): MDAI 2017, LNAI 10571, pp. 15–25, 2017.
DOI: 10.1007/978-3-319-67422-3_2

in the *mean value theorem for integration*. Then it is said that decision making with utility function φ is *more risk averse* than decision making with utility function ψ if $\mu \leq \nu$ for all closed intervals $[a, b]$. Its equivalent condition is

$$\frac{\varphi''}{\varphi'} \leq \frac{\psi''}{\psi'} \tag{1.3}$$

on $\mathbb{R}$ (Yoshida [10,11]).

Yoshida [12] introduced weighted quasi-arithmetic means on two-dimensional regions, which are related to multi-object decision making. In this paper, using decision makers' utility functions we discuss relations between risk averse/risk neutral/risk loving conditions and the corresponding weighted quasi-arithmetic means on two-dimensional regions. In this paper we compare two decision makers' behaviors regarding risks by the weighted quasi-arithmetic means on two-dimensional regions and we give a characterization by their utility functions.

In Sect. 2 we introduce weighted quasi-arithmetic means on two-dimensional regions and we discuss their risk averse conditions. For two utility functions f and g on two-dimensional regions, we introduce a concept that decision making with utility f is more risk averse than decision making with utility g. Further we derive a necessary condition where decision making with utility f is more risk averse than decision making with utility g on two-dimensional regions, and we investigate the condition by several examples. In Sect. 3 we give sufficient conditions for the results in Sect. 2 when utility functions are quadratic.

2   Weighted Quasi-arithmetic Means on Two-Dimensional Regions

Let $\mathbb{R} = (-\infty, \infty)$ and let a domain D be a non-empty open convex subset of $\mathbb{R}^2$, and let $\mathcal{R}(D)$ be a family of closed convex subsets of D. Denote by $\mathcal{L}$ a family of twice continuously differentiable functions $f : D \mapsto \mathbb{R}$ which is strictly increasing, i.e. $f_x > 0$ and $f_y > 0$ on D, and denote by $\mathcal{W}$ a family of continuous functions $w : D \mapsto (0, \infty)$. For a closed convex set $R \in \mathcal{R}(D)$, *weighted quasi-arithmetic means* on region R with utility $f \in \mathcal{L}$ and weighting $w \in \mathcal{W}$ are given by a subset $M_w^f(R)$ of region R as follows.

$$M_w^f(R) = \left\{ (\tilde{x}, \tilde{y}) \in R \mid f(\tilde{x}, \tilde{y}) \iint_R w(x, y)\, dx\, dy = \iint_R f(x, y) w(x, y)\, dx\, dy \right\}. \tag{2.1}$$

Then we have $M_w^f(R) \neq \emptyset$ since f is continuous on R and

$$\min_{(\tilde{x}, \tilde{y}) \in R} f(\tilde{x}, \tilde{y}) \leq \iint_R f(x, y) w(x, y)\, dx\, dy \Big/ \iint_R w(x, y)\, dx\, dy \leq \max_{(\tilde{x}, \tilde{y}) \in R} f(\tilde{x}, \tilde{y}).$$

We introduce the following natural ordering on $\mathbb{R}^2$.

Definition 2.1 (A partial order $\preceq$ on $\mathbb{R}^2$).

(i) For two points $(\underline{x}, \underline{y}), (\overline{x}, \overline{y})(\in \mathbb{R}^2)$, an order $(\underline{x}, \underline{y}) \preceq (\overline{x}, \overline{y})$ implies $\underline{x} \leq \overline{x}$ and $\underline{y} \leq \overline{y}$.

(ii) For two points $(\underline{x}, \underline{y}), (\overline{x}, \overline{y})(\in \mathbb{R}^2)$, an order $(\underline{x}, \underline{y}) \prec (\overline{x}, \overline{y})$ implies $(\underline{x}, \underline{y}) \preceq (\overline{x}, \overline{y})$ and $(\underline{x}, \underline{y}) \neq (\overline{x}, \overline{y})$.

(iii) For two sets $A, B(\subset \mathbb{R}^2)$, an order $A \preceq B$ implies the following (a) and (b):

(a) For any $(\underline{x}, \underline{y}) \in A$ there exists $(\overline{x}, \overline{y}) \in B$ satisfying $(\underline{x}, \underline{y}) \preceq (\overline{x}, \overline{y})$.

(b) For any $(\overline{x}, \overline{y}) \in B$ there exists $(\underline{x}, \underline{y}) \in A$ satisfying $(\underline{x}, \underline{y}) \preceq (\overline{x}, \overline{y})$.

Let a closed convex region $R \in \mathcal{R}(D)$ and let a weighting function $w \in \mathcal{W}$. We define a point $(\overline{x}_R, \overline{y}_R)$ on region R by the following weighted quasi-arithmetic means:

$$\overline{x}_R = \iint_R x\, w(x, y)\, dx\, dy \Big/ \iint_R w(x, y)\, dx\, dy, \qquad (2.2)$$

$$\overline{y}_R = \iint_R y\, w(x, y)\, dx\, dy \Big/ \iint_R w(x, y)\, dx\, dy. \qquad (2.3)$$

Hence, $(\overline{x}_R, \overline{y}_R)$ is called an *invariant risk neutral point on R with weighting w* (Yoshida [12]). We separate the space $\mathbb{R}^2$ as follows. Let $R_{w,-}^{(\overline{x}_R, \overline{y}_R)} = \{(x, y) \in \mathbb{R}^2 \mid (x, y) \prec (\overline{x}_R, \overline{y}_R)\} = \{(x, y) \in \mathbb{R}^2 \mid x \leq \overline{x}_R, y \leq \overline{y}_R, (x, y) \neq (\overline{x}_R, \overline{y}_R)\}$ and $R_{w,+}^{(\overline{x}_R, \overline{y}_R)} = \{(x, y) \in \mathbb{R}^2 \mid (\overline{x}_R, \overline{y}_R) \prec (x, y)\} = \{(x, y) \in \mathbb{R}^2 \mid x \geq \overline{x}_R, y \geq \overline{y}_R, (x, y) \neq (\overline{x}_R, \overline{y}_R)\}$. Then $R_{w,-}^{(\overline{x}_R, \overline{y}_R)}$ denotes a subregion of *risk averse points* and $R_{w,+}^{(\overline{x}_R, \overline{y}_R)}$ denotes a subregion of *risk loving points*. Let $R_w^{(\overline{x}_R, \overline{y}_R)} = R_{w,-}^{(\overline{x}_R, \overline{y}_R)} \cup R_{w,+}^{(\overline{x}_R, \overline{y}_R)} \cup \{(\overline{x}_R, \overline{y}_R)\}$. Now we introduce the following relations between decision maker's behavior and his utility.

Definition 2.2. Let a utility function $f \in \mathcal{L}$ and let a rectangle region $R \in \mathcal{R}(D)$.

(i) Decision making with utility f is called *risk neutral on R* if

$$f(\overline{x}_R, \overline{y}_R) \iint_R w(x, y)\, dx\, dy = \iint_R f(x, y) w(x, y)\, dx\, dy \qquad (2.4)$$

for all density functions w.

(ii) Decision making with utility f is called *risk averse on R* if

$$f(\overline{x}_R, \overline{y}_R) \iint_R w(x, y)\, dx\, dy \geq \iint_R f(x, y) w(x, y)\, dx\, dy \qquad (2.5)$$

for all density functions w.

(iii) Decision making with utility f is called *risk loving on R* if

$$f(\overline{x}_R, \overline{y}_R) \iint_R w(x, y)\, dx\, dy \leq \iint_R f(x, y) w(x, y)\, dx\, dy \qquad (2.6)$$

for all density functions w.

Example 2.1. Let a domain $D = (-0.5, 1.25)^2$ and a region $R = [0, 1]^2$, and let a weighting function $w(x, y) = 1$ for $(x, y) \in D$. Then an invariant neutral point is $(\overline{x}_R, \overline{y}_R) = (0.5, 0.5)$ and $R_{w,-}^{(\overline{x}_R, \overline{y}_R)} = [0, 0.5]^2 \setminus \{(0.5, 0.5)\}$ and $R_{w,+}^{(\overline{x}_R, \overline{y}_R)} = [0.5, 1]^2 \setminus \{(0.5, 0.5)\}$. Let us consider two utility functions $f(x, y) = -x^2 - y^2 + 3x + 3y$ and $g(x, y) = 2x^2 + 2y^2 - 5x - 5y$ for $(x, y) \in D$. Then by Yoshida [12, Example 3.1(i), Lemma 2.2] decision making with utility function f is called risk averse on R with weighting w, and decision making with utility function g is also called risk loving on R with weighting w. Hence the corresponding weighted quasi-arithmetic means $M_w^f(R)$ and $M_w^g(R)$ are ordered by the order $\preceq$ in a restricted subregion $R_w^{(\overline{x}_R, \overline{y}_R)} = R_{w,-}^{(\overline{x}_R, \overline{y}_R)} \cup R_{w,+}^{(\overline{x}_R, \overline{y}_R)} \cup \{(\overline{x}_R, \overline{y}_R)\}$. However they can not be ordered on a subregion $R \setminus R_w^{(\overline{x}_R, \overline{y}_R)}$ (Fig. 1).

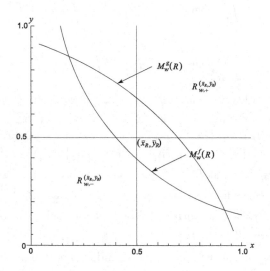

Fig. 1. $M_w^f(R) \cap R_w^{(\overline{x}_R, \overline{y}_R)} \preceq M_v^g(R) \cap R_w^{(\overline{x}_R, \overline{y}_R)}$ ($f(x, y) = -x^2 - y^2 + 3x + 3y, g(x, y) = 2x^2 + 2y^2 - 5x - 5y$, $R = [0, 1]^2$)

It is natural that the order $\preceq$ should be given between weighted quasi-arithmetic means $M_w^f(R)$ of risk averse utility f and weighted quasi-arithmetic means $M_w^g(R)$ of risk loving utility g in Example 3.1. Therefore when we compare weighted quasi-arithmetic means $M_w^f(R)$ and $M_v^g(R)$, we discuss it on the meaningful restricted subregion $R_w^{(\overline{x}_R, \overline{y}_R)}$. Hence we introduce the following definition regarding the comparison of utility functions.

Definition 2.3. Let $f, g \in \mathcal{L}$ be utility functions on D. Decision making with utility f is *more risk averse than* decision making with utility g if it holds that

$$M_w^f(R) \cap R_w^{(\overline{x}_R, \overline{y}_R)} \preceq M_v^g(R) \cap R_w^{(\overline{x}_R, \overline{y}_R)} \tag{2.7}$$

for all weighting functions $w \in \mathcal{W}$ on D and all closed convex regions $R \in \mathcal{R}(D)$.

Example 2.2. Let a domain $D = (-0.5, 1.25)^2$ and a region $R = [0, 1]^2$, and let a weighting function $w(x, y) = 1$ for $(x, y) \in D$. Then an invariant neutral point is $(\overline{x}_R, \overline{y}_R) = (0.5, 0.5)$ and $R_{w,-}^{(\overline{x}_R, \overline{y}_R)} = [0, 0.5]^2 \setminus \{(0.5, 0.5)\}$ and $R_{w,+}^{(\overline{x}_R, \overline{y}_R)} = [0.5, 1]^2 \setminus \{(0.5, 0.5)\}$. Let us consider two utility functions $f(x, y) = -x^2 - y^2 + 3x + 3y$ and $g(x, y) = -2x^2 - 2y^2 + 5x + 5y$ for $(x, y) \in D$. Then decision making with utility f is *more risk averse than* decision making with utility g as we see the relation (2.7) in Fig. 2.

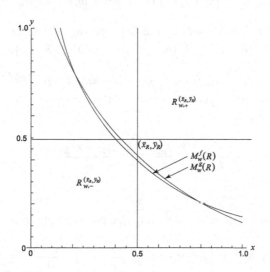

Fig. 2. $M_w^f(R) \cap R_w^{(\overline{x}_R, \overline{y}_R)} \preceq M_v^g(R) \cap R_w^{(\overline{x}_R, \overline{y}_R)}$ $(f(x, y) = -x^2 - y^2 + 3x + 3y, g(x, y) = -2x^2 - 2y^2 + 5x + 5y, R = [0, 1]^2)$

Now we give a necessary condition for (2.7), i.e. decision making with utility f is more risk averse than decision making with utility g.

Theorem 2.1. *Let $f, g \in \mathcal{L}$ be utility functions on D. If decision making with utility f is more risk averse than decision making with utility g, then it holds that*

$$\frac{h^2 f_{xx} + 2rhk f_{xy} + k^2 f_{yy}}{h f_x + k f_y} \leq \frac{h^2 g_{xx} + 2rhk g_{xy} + k^2 g_{yy}}{h g_x + k g_y} \qquad (2.8)$$

on D for all positive numbers h and k and all real numbers r satisfying $-1 \leq r \leq 1$.

From Theorem 2.1 we can easily obtain the following result, which is corresponding to [12, Theorem 3.1(i)].

Corollary 2.1. *Let $f, g \in \mathcal{L}$ be utility functions on D. If decision making with utility f is more risk averse than decision making with utility g, then it holds that*

$$\frac{f_{xx}}{f_x} \leq \frac{g_{xx}}{g_x} \quad and \quad \frac{f_{yy}}{f_y} \leq \frac{g_{yy}}{g_y} \quad on \; D. \qquad (2.9)$$

Equation (2.8) in Theorem 2.1 gives a detailed relation between f and g rather than (2.9). A parameter r in necessary condition (2.8) depends on the shapes of closed convex regions $R \in \mathcal{R}(D)$. Now we investigate several examples with different shapes of regions R.

Example 2.3 (Rectangle regions). Let h and k be positive numbers. Let rectangle regions

$$R_{h,k}^{\text{Rect}}(a, b, t) = [a, a + ht] \times [b, b + kt] \qquad (2.10)$$

for $(a, b) \in D$ and $t > 0$. Denote a family of rectangle regions by $\mathcal{R}_{h,k}^{\text{Rect}}(D) = \{R_{h,k}^{\text{Rect}}(a, b, t) \mid R_{h,k}^{\text{Rect}}(a, b, t) \subset D, (a, b) \in D, t > 0\} (\subset \mathcal{R}(D))$, (Fig. 3).

Corollary 2.2. *If utility functions* $f, g \in \mathcal{L}$ *satisfy* $M_w^f(R) \cap R_w^{(\overline{x}_R, \overline{y}_R)} \preceq M_v^g(R) \cap R_w^{(\overline{x}_R, \overline{y}_R)}$ *for all weighting functions* $w \in \mathcal{W}$ *on* D *and all rectangle regions* $R \in \mathcal{R}_{h,k}^{\text{Rect}}(D)$, *then it holds that*

$$\frac{h^2 f_{xx} + k^2 f_{yy}}{h f_x + k f_y} \leq \frac{h^2 g_{xx} + k^2 g_{yy}}{h g_x + k g_y} \qquad (2.11)$$

on D.

Example 2.4 (Oval regions). Let h and k be positive numbers. Let oval regions

$$R_{h,k}^{\text{Oval}}(a, b, t) = \left\{ (x, y) \in \mathbb{R}^2 \,\middle|\, \frac{(x - a)^2}{h^2} + \frac{(y - b)^2}{k^2} \leq t^2 \right\} \qquad (2.12)$$

for $(a, b) \in D$ and $t > 0$. Denote a family of oval regions by $\mathcal{R}_{h,k}^{\text{Oval}}(D) = \{R_{h,k}^{\text{Oval}}(a, b, t) \mid R_{h,k}^{\text{Oval}}(a, b, t) \subset D, (a, b) \in D, t > 0\} (\subset \mathcal{R}(D))$, (Fig. 3).

Corollary 2.3. *If utility functions* $f, g \in \mathcal{L}$ *satisfy* $M_w^f(R) \cap R_w^{(\overline{x}_R, \overline{y}_R)} \preceq M_v^g(R) \cap R_w^{(\overline{x}_R, \overline{y}_R)}$ *for all weighting functions* $w \in \mathcal{W}$ *on* D *and all oval regions* $R \in \mathcal{R}_{h,k}^{\text{Oval}}(D)$, *then it holds that*

$$\frac{h^2 f_{xx} + k^2 f_{yy}}{h f_x + k f_y} \leq \frac{h^2 g_{xx} + k^2 g_{yy}}{h g_x + k g_y} \qquad (2.13)$$

on D.

Example 2.5 (Triangle regions). Let h and k be positive numbers. Let triangle regions

$$R_{h,k}^{\text{Tri}}(a, b, t) = \left\{ (x, y) \in \mathbb{R}^2 \,\middle|\, x \geq a, \, y \geq b, \, \frac{x - a}{h} + \frac{y - b}{k} \leq t \right\} \qquad (2.14)$$

for $(a, b) \in D$ and $t > 0$. Denote a family of triangle regions by $\mathcal{R}_{h,k}^{\text{Tri}}(D) = \{R_{h,k}^{\text{Tri}}(a, b, t) \mid R_{h,k}^{\text{Tri}}(a, b, t) \subset D, (a, b) \in D, t > 0\} (\subset \mathcal{R}(D))$, (Fig. 4).

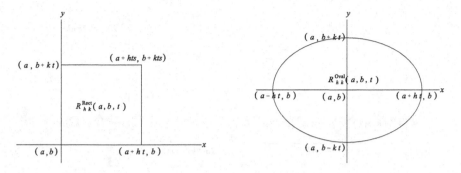

Fig. 3. Rectangle region $R_{h,k}^{\mathrm{Rect}}(a,b,t)$ and oval region $R_{h,k}^{\mathrm{Oval}}(a,b,t)$

Corollary 2.4. *If utility functions* $f, g \in \mathcal{L}$ *satisfy* $M_w^f(R) \cap R_w^{(\overline{x}_R, \overline{y}_R)} \preceq M_v^g(R) \cap$ $R_w^{(\overline{x}_R, \overline{y}_R)}$ *for all weighting functions* $w \in \mathcal{W}$ *on* D *and all triangle regions* $R \in$ $\mathcal{R}_{h,k}^{\mathrm{Tri}}(D)$*, then it holds that*

$$\frac{h^2 f_{xx} - hk f_{xy} + k^2 f_{yy}}{h f_x + k f_y} \leq \frac{h^2 g_{xx} - hk g_{xy} + k^2 g_{yy}}{h g_x + k g_y} \tag{2.15}$$

on D.

Example 2.6 (Parallelogram regions)**.** Let h and k be positive numbers. Let parallelogram regions

$$R_{h,k}^{\mathrm{Para}}(a,b,t) = \{(x,y) \mid |k(x-a) - 3h(y-b)| \leq 4hkt, \ |3k(x-a) - h(y-b)| \leq 4hkt\} \tag{2.16}$$

for $(a,b) \in D$ and $t > 0$. Denote a family of parallelogram regions by $\mathcal{R}_{h,k}^{\mathrm{Para}}(D) = \{R_{h,k}^{\mathrm{Para}}(a,b,t) \mid R_{h,k}^{\mathrm{Para}}(a,b,t) \subset D, \ (a,b) \in D, \ t > 0\}(\subset \mathcal{R}(D))$, (Fig. 4).

Corollary 2.5. *If utility functions* $f, g \in \mathcal{L}$ *satisfy* $M_w^f(R) \cap R_w^{(\overline{x}_R, \overline{y}_R)} \preceq M_v^g(R) \cap$ $R_w^{(\overline{x}_R, \overline{y}_R)}$ *for all weighting functions* $w \in \mathcal{W}$ *on* D *and all parallelogram regions* $R \in \mathcal{R}_{h,k}^{\mathrm{Para}}(D)$*, then it holds that*

$$\frac{h^2 f_{xx} + \frac{3}{5} hk f_{xy} + k^2 f_{yy}}{h f_x + k f_y} \leq \frac{h^2 g_{xx} + \frac{3}{5} hk g_{xy} + k^2 g_{yy}}{h g_x + k g_y} \tag{2.17}$$

on D.

Example 2.3 (Rectangle regions) and Example 2.4 (Oval regions) are cases where $r = 0$ in (2.8), and Example 2.5 (Triangle regions) and Example 2.6 (Parallelogram regions) are cases where $r = -\frac{1}{2}$ and $r = \frac{3}{10}$ respectively in (2.8).

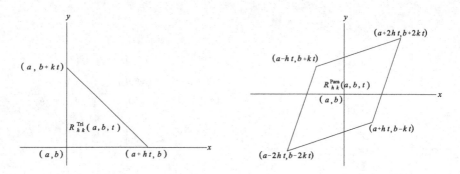

Fig. 4. Triangle region $R_{h,k}^{\mathrm{Tri}}(a,b,t)$ and parallelogram region $R_{h,k}^{\mathrm{Para}}(a,b,t)$

3   A Sufficient Condition

Let $f,g \in \mathcal{L}$ be utility functions on an open convex domain D. Theorem 2.1 gives a necessary condition that decision making with utility f is more risk averse than decision making with utility g. In this section, we discuss its sufficient condition. For a utility function $f \in \mathcal{L}$, its Hessian matrix is written by

$$H^f(x,y) = \begin{pmatrix} f_{xx}(x,y) & f_{xy}(x,y) \\ f_{yx}(x,y) & f_{yy}(x,y) \end{pmatrix} \tag{3.1}$$

for $(x,y) \in D$. The the following proposition gives a sufficient condition for (2.8) in Theorem 2.1.

Proposition 3.1. *Let $f,g \in \mathcal{L}$ be utility functions on D. Then the following (i) and (ii) hold.*

(i) Matrices

$$\frac{1}{f_x(x,y)}H^f(x,y) - \frac{1}{g_x(x,y)}H^g(x,y) \text{ and } \frac{1}{f_y(x,y)}H^f(x,y) - \frac{1}{g_y(x,y)}H^g(x,y) \tag{3.2}$$

are negative semi-definite for all $(x,y) \in D$ if and only if a matrix

$$\frac{1}{hf_x(x,y) + kf_y(x,y)}H^f(x,y) - \frac{1}{hg_x(x,y) + kg_y(x,y)}H^g(x,y) \tag{3.3}$$

is negative semi-definite for all $(x,y) \in D$ and all positive numbers h and k.
(ii) If (3.2) are negative semi-definite at all $(x,y) \in D$, then (2.8) holds on D for all positive numbers h and k and all real numbers r satisfying $-1 \le r \le 1$.

From Proposition 3.1 implies that the condition (3.2) is stronger than the condition (2.8), however (3.2) is easier than (2.8) to check in actual cases. In this paper, utility functions $f(\in \mathcal{L})$ are called *quadratic* if the second derivatives

f_{xx}, f_{xy} and f_{yy} are constant functions. When utility functions are quadratic, the following theorem gives a sufficient condition for what decision making with utility f is more risk averse than decision making with utility g.

Theorem 3.1. *Let utility functions* $f, g \in \mathcal{L}$ *be quadratic on D. If*

$$\frac{1}{f_x(x,y)}H^f(x,y) - \frac{1}{g_x(x,y)}H^g(x,y) \quad and \quad \frac{1}{f_y(x,y)}H^f(x,y) - \frac{1}{g_y(x,y)}H^g(x,y) \quad (3.4)$$

are negative semi-definite at all $(x,y) \in D$, *then decision making with utility* f *is more risk averse than decision making with utility* g, *i.e.*

$$M_w^f(R) \cap R_w^{(\overline{x}_R, \overline{y}_R)} \preceq M_v^g(R) \cap R_w^{(\overline{x}_R, \overline{y}_R)}$$

for all weighting functions $w \in \mathcal{W}$ *and all closed convex regions* $R \in \mathcal{R}(D)$.

Now we give an example for Theorem 3.1.

Example 3.1 (Quadratic utility functions). Let a domain $D = (-0.5, 1.5)^2$ and a region $R = [0,1]^2$, and let a weighting function $w(x,y) = 1$ for $(x,y) \in D$. Then an invariant neutral point is $(\overline{x}_R, \overline{y}_R) = (0.5, 0.5)$ and $R_{w,-}^{(\overline{x}_R, \overline{y}_R)} = [0, 0.5]^2 \setminus \{(0.5, 0.5)\}$ and $R_{w,+}^{(\overline{x}_R, \overline{y}_R)} = [0.5, 1]^2 \setminus \{(0.5, 0.5)\}$. Let us consider two quadratic utility functions $f(x,y) = -2x^2 - 2y^2 + 2xy + 8x + 8y$ and $g(x,y) = -x^2 - y^2 + xy + 5x + 5y$ for $(x,y) \in D$. Then f and g are increasing on D, i.e. $f_x(x,y) = -4x + 2y + 8 > 0$, $f_y(x,y) = 2x - 4y + 8 > 0$, $g_x(x,y) = -2x + y + 5 > 0$ and $g_y(x,y) = x - 2y + 5 > 0$ on D. Their Hessian matrices are

$$H^f(x,y) = \begin{pmatrix} -4 & 2 \\ 2 & -4 \end{pmatrix} \quad and \quad H^g(x,y) = \begin{pmatrix} -2 & 1 \\ 1 & -2 \end{pmatrix}. \quad (3.5)$$

Let $A(x,y)$ and $B(x,y)$ by $A(x,y) = \frac{1}{f_x(x,y)}H^f(x,y) - \frac{1}{g_x(x,y)}H^g(x,y)$ and $B(x,y) = \frac{1}{f_y(x,y)}H^f(x,y) - \frac{1}{g_y(x,y)}H^g(x,y)$ for $(x,y) \in D$, and then we have

$$A(x,y) = \frac{1}{-4x + 2y + 8}\begin{pmatrix} -4 & 2 \\ 2 & -4 \end{pmatrix} - \frac{1}{-2x + y + 5}\begin{pmatrix} -2 & 1 \\ 1 & -2 \end{pmatrix}, \quad (3.6)$$

$$B(x,y) = \frac{1}{2x - 4y + 8}\begin{pmatrix} -4 & 2 \\ 2 & -4 \end{pmatrix} - \frac{1}{x - 2y + 5}\begin{pmatrix} -2 & 1 \\ 1 & -2 \end{pmatrix}. \quad (3.7)$$

We can easily check $A(x,y)$ and $B(x,y)$ are negative definite for all $(x,y) \in D$. From Theorem 3.1, decision making with utility f is more risk averse than decision making with utility g on R and it holds that $M_w^f(R) \cap R_w^{(\overline{x}_R, \overline{y}_R)} \preceq M_v^g(R) \cap R_w^{(\overline{x}_R, \overline{y}_R)}$ for all weighting functions $w \in \mathcal{W}$ (Fig. 5).

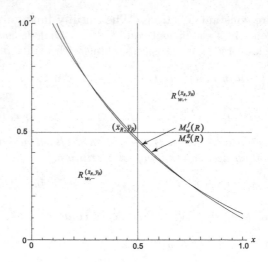

Fig. 5. $M_w^f(R) \cap R_w^{(\bar{x}_R, \bar{y}_R)} \preceq M_v^g(R) \cap R_w^{(\bar{x}_R, \bar{y}_R)}$ $(f(x,y) = -2x^2 - 2y^2 + 2xy + 8x + 8y, g(x,y) = -x^2 - y^2 + xy + 5x + 5y, R = [0,1]^2)$

Concluding Remark. When utility functions are quadratic, Theorem 3.1 gives a sufficient condition where decision making with utility f is more risk averse than decision making with utility g. It is an open problem whether (3.2) is a sufficient condition when utility functions are not quadratic but more general.

Acknowledgments. This research is supported from JSPS KAKENHI Grant Number JP 16K05282.

References

1. Aczél, J.: On weighted mean values. Bulletin of the American Math. Society **54**, 392–400 (1948)
2. Bustince, H., Calvo, T., Baets, B., Fodor, J., Mesiar, R., Montero, J., Paternain, D., Pradera, A.: A class of aggregation functions encompassing two-dimensional OWA operators. Inf. Sci. **180**, 1977–1989 (2010)
3. Fishburn, P.C.: Utility Theory for Decision Making. Wiley, New York (1970)
4. Kolmogoroff, A.N.: Sur la notion de la moyenne. Acad. Naz. Lincei Mem. Cl. Sci. Fis. Mat. Natur. Sez. **12**, 388–391 (1930)
5. Labreuche, C., Grabisch, M.: The Choquet integral for the aggregation of interval scales in multicriteria decision making. Fuzzy Sets Syst. **137**, 11–26 (2003)
6. Nagumo, K.: Über eine Klasse der Mittelwerte. Japan. J. Math. **6**, 71–79 (1930)
7. Torra, V., Godo, L.: On defuzzification with continuous WOWA operators. In: Aggregation Operators, pp. 159–176. Springer, New York (2002)
8. Yoshida, Y.: Aggregated mean ratios of an interval induced from aggregation operations. In: Torra, V., Narukawa, Y. (eds.) MDAI 2008. LNCS, vol. 5285, pp. 26–37. Springer, Heidelberg (2008). doi:10.1007/978-3-540-88269-5_4

9. Yoshida, Y.: Quasi-arithmetic means and ratios of an interval induced from weighted aggregation operations. Soft Comput. **14**, 473–485 (2010)

10. Yoshida, Y.: Weighted quasi-arithmetic means and conditional expectations. In: Torra, V., Narukawa, Y., Daumas, M. (eds.) MDAI 2010. LNCS, vol. 6408, pp. 31–42. Springer, Heidelberg (2010). doi:10.1007/978-3-642-16292-3_6

11. Yoshida, Y.: Weighted quasi-arithmetic means and a risk index for stochastic environments. Int. J. Uncertainty Fuzziness Knowl. Based Syst. **16**(suppl), 1–16 (2011)

12. Yoshida, Y.: Weighted quasi-arithmetic means on two-dimensional regions and their applications. In: Torra, V., Narukawa, Y. (eds.) MDAI 2015. LNCS, vol. 9321, pp. 42–53. Springer, Cham (2015). doi:10.1007/978-3-319-23240-9_4

Symmetrization Methods for Aggregation Functions

Radko Mesiar and Andrea Stupňanová[✉]

Faculty of Civil Engineering, Slovak University of Technology in Bratislava,
Bratislava, Slovak Republic
{radko.mesiar,andrea.stupnanova}@stuba.sk

Abstract. We introduce and discuss the concept of symmetrization methods for aggregation functions. Several symmetrization methods are exemplified. A particular stress is put on extremal symmetrization methods.

Keywords: Aggregation function · Anonymity · Permutation · Symmetric aggregation function · Symetrization method

1 Introduction

Symmetry of aggregation functions can be seen as a generalization of the commutativity of binary operations $x * y = y * x$, and it is known also as neutrality or anonymity. This important property indicates the equal treating of any considered input to be aggregated. It is crucial in any situation when the order of considered inputs is not known, for example when the inputs to be aggregated are evaluations of jury members stored after an anonymous procedure in a voting box (this example was motivating to call this property as anonymity in the field of multicriteria decision support). Two distinguished symmetrization methods, i.e., methods relating to a considered aggregation function some symmetric aggregation functions, can be found in [4], see also [5]. Namely, considering an n-ary real function F and an n-ary input vector $\mathbf{x} = (x_1, \ldots x_n)$, the function F^+ and F^- given by

$$F^+(\mathbf{x}) = F(x_{(1)}, \ldots, x_{(n)}) \quad \text{and} \quad F^-(\mathbf{x}) = F(x_{(n)}, \ldots, x_{(1)}),$$

where $(.)$ is a permutation such that $x_{(1)} \leq \cdots \leq x_{(n)}$, are symmetric. A generalization of these symmetrization methods based on a fixed permutation $\sigma \in \mathcal{P}_n$ (the set of all permutations on $\{1, \ldots, n\}$) was introduced in [5], proposing a function $F_{(\sigma)}$ given by

$$F_{(\sigma)} = F(x_{(\sigma(1))}, \ldots, x_{(\sigma(n))}).$$

Obviously $F_{(id)} = F^+$ and $F_{(rev)} = F^-$, where $id, rev \in \mathcal{P}_n$ are the identity permutation $(1, \ldots, n)$ and the reversed permutation $(n, \ldots, 1)$, respectively.

© Springer International Publishing AG 2017
V. Torra et al. (Eds.): MDAI 2017, LNAI 10571, pp. 26–32, 2017.
DOI: 10.1007/978-3-319-67422-3_3

Observe that the permutation (.) depends on the input vector $\mathbf{x}$ and it need not be unique (this happens if there are some ties between the arguments $x_1, \ldots, x_n$). However, this possible non-uniqueness does not influence the fact that F_σ is well defined and symmetric for any permutation $\sigma \in \mathcal{P}_n$ (σ is independent of any considered input vector $\mathbf{x}$).

The aim of this contribution is to introduce and discuss an axiomatic approach to symmetrization of aggregation functions. The paper is organized as follows. In the next section, the necessary preliminaries are given. Section 3 brings our axiomatic characterization of symmetrization methods and offers several examples of symmetrization methods. In particular, two extremal symmetrization methods are described. In Sect. 4 we apply some of the introduced symmetrization methods to some non-symmetric aggregation functions, and especially to weighted arithmetic means. Some interesting observations are added in the concluding remarks.

2 . Preliminaries

For a fixed $n \geq 1$, a mapping $A : [0,1]^n \to [0,1]$ is called an aggregation function whenever it is increasing in each coordinate and it satisfies the boundary conditions $A(\mathbf{0}) = A(0, \ldots, 0) = 0$ and $A(\mathbf{1}) = A(1, \ldots, 1) = 1$. Note that viewing $[0,1]^n$ and $[0,1]$ as bounded lattices, aggregation functions are just the order-homomorphisms.

The class of all n-ary aggregation functions is denoted as $\mathcal{A}_n$. Equipped with the partial order of n-ary real functions, $\mathcal{A}_n$ is a bounded lattice with the top element $A_\top$ and the bottom element $A_\bot$, given respectively by

$$A_\top(\mathbf{x}) = \begin{cases} 0 & \text{if } \mathbf{x} = \mathbf{0} \\ 1 & \text{otherwise} \end{cases} \quad \text{and} \quad A_\bot(\mathbf{x}) = \begin{cases} 1 & \text{if } \mathbf{x} = \mathbf{1} \\ 0 & \text{otherwise} \end{cases}.$$

For more details concerning aggregation functions we recommend monographs [1,2,5].

An aggregation function $A \in \mathcal{A}_n$ is called symmetric whenever

$$A(\mathbf{x}) = A(\mathbf{x}_\sigma) \quad \text{for any } \mathbf{x} \in [0,1]^n$$

and any permutation $\sigma \in \mathcal{P}_n$, where $\mathbf{x}_\sigma = (x_{\sigma(1)}, \ldots, x_{\sigma(n)})$.
Observe that the class $\mathcal{P}_n$ can be generated by two permutations, say σ_1 and σ_2 (i.e., any $\sigma \in \mathcal{P}_n$ can be obtained from σ_1 and σ_2, applying the composition operator consecutively), and then the symmetry of an aggregation function A is characterized by the equality

$$A(\mathbf{x}) = A(\mathbf{x}_{\sigma_1}) = A(\mathbf{x}_{\sigma_2}) \quad \text{valid for each } \mathbf{x} \in [0,1]^n.$$

As one example recall $\sigma_1 = (2, 1, 3, \ldots, n)$ and $\sigma_2 = (2, 3, \ldots, n, 1)$.

The class of all n-ary symmetric aggregation functions is denoted as $\mathcal{A}_{ns}$. It is evident that $\mathcal{A}_{ns}$ is a sublattice of $\mathcal{A}_n$ with the top element $A_\top$ and the bottom element $A_\bot$. Both classes $\mathcal{A}_n$ and $\mathcal{A}_{ns}$ are closed under composition by means of $B \in \mathcal{A}_k$, i.e.,

○ for any $k \geq 2$, $B \in \mathcal{A}_k$ and $A_1, \ldots, A_k \in \mathcal{A}_n$ $(A_1, \ldots, A_k \in \mathcal{A}_{ns})$ also the composite $B(A_1, \ldots, A_k) \in \mathcal{A}_n$ $(B(A_1, \ldots, A_k) \in \mathcal{A}_{ns})$.

Among several other properties of aggregation functions discussed in [1,2,5], we recall the idempotency. An aggregation function $A \in \mathcal{A}_n$ is called idempotent (averaging, compensative) whenever

$$A(\mathbf{c}) = A(c, \ldots, c) \quad \text{for any constant } c \in [0, 1].$$

Equivalently, the idempotency of aggregation functions can be characterized by the averaging property

$$\min(x_1, \ldots, x_n) \leq A(\mathbf{x}) \leq \max(x_1, \ldots, x_n).$$

3   Symmetrization Methods for Aggregation Functions

Any symmetrization method for (n-ary) aggregation functions should assign to an aggregation function $A \in \mathcal{A}_n$ some idempotent aggregation function $A^s \in \mathcal{A}_{ns}$. We expect that any such reasonable symmetrization method

- does not change the symmetric aggregation functions, i.e.,

$$A = A^s \quad \text{whenever} \quad A \in \mathcal{A}_{ns}, \text{ and}$$

- preserves the ordering of aggregation functions, i.e.,

$$\text{if } A, B \in \mathcal{A}_n, A \leq B, \quad \text{then} \quad A^s \leq B^s.$$

Formally, we propose the next axiomatic approach to symmetrization of aggregation functions.

Definition 1. *A mapping $\varphi : \mathcal{A}_n \to \mathcal{A}_{ns}$ is called a symmetrization method (for n-ary aggregation functions) whenever it is simultaneously*

(i) an order homomorphism;
(ii) a projection.

Hence, $\varphi : \mathcal{A}_n \to \mathcal{A}_{ns}$ is a symmetrization method whenever

$A, B \in \mathcal{A}_n, A \leq B$, implies $\varphi(A) \leq \varphi(B)$, and $\varphi(A) = A$ for any $A \in \mathcal{A}_{ns}$.

Clearly, then $\varphi(\varphi(A)) = \varphi(A)$. All till now mentioned symmetrization methods (recall F^+, F^- and $F_{(\sigma)}$) satisfy Definition 1. They are based on the permutation $(.) \in \mathcal{P}_n$ which depends on $\mathbf{x}$. This observation allows to split the class of all symmetrization methods into two subclasses:

- input dependent symmetrization methods;
- input independent symmetrization methods.

Note that though the input vector $\mathbf{x}$ is necessarily considered when processing $\varphi(A)(\mathbf{x})$ for any symmetrization method φ, the above partition indicates whether the introduction of φ requires the determination of $\mathbf{x}$-dependent permutation (.) or not. The next results allow to introduce a rich variety of symmetrization methods of both kinds.

Theorem 1. *For a fixed $k \geq 1$, for any idempotent aggregation function $B \in \mathcal{A}_k$ and k-tuple $K = (\sigma_1, \ldots, \sigma_k) \in (\mathcal{P}_n)^k$ of permutations, the mapping $\varphi : \mathcal{A}_n \to \mathcal{A}_{ns}$ given by*

$$\varphi(A) = A_{B,K} = B(A_{(\sigma_1)}, \ldots, A_{(\sigma_k)}), \ i.e.,$$
$$A_{B,K}(\mathbf{x}) = B(A(\mathbf{x}_{(\sigma_1)}), \ldots, A(\mathbf{x}_{(\sigma_k)}))$$

is an input dependent symmetrization method.

Proof.

(i) Note that if $k = 1$ then $B(x) = x, x \in [0,1]$, and $K = (\sigma) \in \mathcal{P}_n$, and thus $A_{B,K} = A_{(\sigma)}$.

(ii) For $k \geq 2$, the symmetry of $\varphi(A)$ was discussed in Sect. 2. Moreover, if $A \in \mathcal{A}_{ns}$ then $A_{(\sigma)} = A$ for any $\sigma \in \mathcal{P}_n$. Then the idempotency of B ensures $\varphi(A) = B(A_{(\sigma_1)}, \ldots, A_{(\sigma_k)}) = B(A, \ldots, A) = A$, thus proving that φ is a symmetrization method. Clearly, it is input dependent. □

To illustrate Theorem 1, consider $n = 3, k = 2, K = (id, rev)$ and $B \in \mathcal{A}_2$ given by $B(x_1, x_2) = \frac{x_1 + 2x_2}{3}$. Then, for any $A \in \mathcal{A}_3$,

$$A_{B,K}(x_1, x_2, x_3) = \frac{A(x_{(1)}, x_{(2)}, x_{(3)}) + 2A(x_{(3)}, x_{(2)}, x_{(1)})}{3}.$$

Suppose $A(x_1, x_2, x_3) = \sqrt[6]{x_1 x_2{}^2 x_3{}^3}$ (i.e., A is a weighted geometric mean). Then

$$A_{B,K}(x_1, x_2, x_3) = \sqrt[6]{x_1 x_2 x_3} \frac{2\sqrt[6]{x_{(1)}} + \sqrt[3]{x_{(3)}}}{3} \sqrt[6]{x_{(2)}}$$
$$= \sqrt[6]{x_1 x_2 x_3} \frac{2\sqrt[3]{\min(x_1, x_2, x_3)} + \sqrt[3]{\max(x_1, x_2, x_3)}}{3} \sqrt[6]{\mathrm{med}(x_1, x_2, x_3)}.$$

The symmetry of $A_{B,K}$ is obvious.

Theorem 2. *Let $B \in \mathcal{A}_{n!}$ be an idempotent symmetric aggregation function of dimension $n!$. Then the mapping $\varphi : \mathcal{A}_n \to \mathcal{A}_{ns}$ given by $\varphi(A) = A_B$,*

$$A_B(\mathbf{x}) = B(A(\mathbf{x}_\sigma) | \sigma \in \mathcal{P}_n),$$

where $\mathbf{x}_\sigma = (x_{\sigma(1)}, \ldots, x_{\sigma(n)})$, and A_B is the B-aggregation of all $n!$ values $A(\mathbf{x}_\sigma), \sigma \in \mathcal{P}_n$, is an input independent symmetrization method.

Proof. Evidently, A_B is an n-ary aggregation function not dependent on **x**-dependent permutation (.). Moreover, due to the facts that for any permutation $\tau \in \mathcal{P}_n$, it holds

$$\{\sigma \circ \tau | \sigma \in \mathcal{P}_n\} = \mathcal{P}_n, \quad \text{and} \quad (\mathbf{x}_\tau)_\sigma = \mathbf{x}_{\sigma \circ \tau},$$

we have

$$A_B(\mathbf{x}_\tau) = B(A(\mathbf{x}_{\sigma \circ \tau}) | \sigma \in \mathcal{P}_n) = B(A(\mathbf{x}_\sigma) | \sigma \in \mathcal{P}_n) = A_B(\mathbf{x}).$$

Hence A_B is symmetric. Finally, if $A \in \mathcal{A}_{ns}$, the idempotency of B ensures $A_B = A$. Summarizing, we have shown that φ is an input independent symmetrization method. □

It is evident that that if $B_1, B_2 \in \mathcal{A}_{n!}, B_1 \leq B_2$, then also $A_{B_1} \leq A_{B_2}$ for any $A \in \mathcal{A}_n$.

Recall that the greatest idempotent aggregation function is the max operator, while the smallest one is the min operator. Moreover, both these functions are symmetric. These facts indicate the next interesting result.

Theorem 3. *Denote* $\varphi^*(A) = A^* = A_{\max}$ *and* $\varphi_*(A) = A_* = A_{\min}$. *Then for any symmetrization method* φ *and any* $A \in \mathcal{A}_n$ *it holds*

$$\varphi_*(A) \leq \varphi(A) \leq \varphi^*(A),$$

i.e., φ^* *is the greatest symmetrization method and* φ_* *is the smallest symmetrization method.*

Proof. Recall that $A(\mathbf{x}_{id}) = A(\mathbf{x})$ and thus $A_* \leq A \leq A^*$. Due to the preservation of order of any symmetrization method φ it holds

$$\varphi_*(A) = A_* = \varphi(A_*) \leq \varphi(A) \leq \varphi(A^*) = A^* = \varphi^*(A). \qquad □$$

Note that extremal symmetrizations A_* and A^* were introduced and discussed in our recent paper [6].

As an interesting input independent symmetrization method we recall that the arithmetic mean AM satisfies all constraints of Theorem 3, and then

$$A_{\mathrm{AM}}(\mathbf{x}) = \frac{1}{n!} \sum_{\sigma \in \mathcal{P}_n} A(\mathbf{x}_\sigma).$$

4   Examples

As a prototypical aggregation function which is not symmetric we recall the weighted arithmetic mean $W_\mathbf{w} : [0,1]^n \to [0,1]$ given by

$$W_\mathbf{w}(\mathbf{x}) = \sum_{i=1}^n w_i x_i,$$

where the weighting vector $\mathbf{w} \in [0,1]^n$ satisfies $\sum_{i=1}^{n} w_i = 1$ and $\mathbf{w} \neq \left(\frac{1}{n}, \ldots, \frac{1}{n}\right)$ (obviously, if $\mathbf{w} = \left(\frac{1}{n}, \ldots, \frac{1}{n}\right)$ then $W_{\mathbf{w}} = $ AM is the arithmetic mean which is a symmetric aggregation function). Then, applying different symmetrization methods, it holds:

- $(W_{\mathbf{w}})_{\text{AM}} = $ AM;
- $(W_{\mathbf{w}})^* = \text{OWA}_{\mathbf{w}^*}$ is the OWA operator [7], where $\mathbf{w}^* = (w_{[1]}, \ldots, w_{[n]})$, $[.] \in \mathcal{P}_n$ being a permutation such that $w_{[1]} \leq w_{[2]} \leq \cdots \leq w_{[n]}$, and then

$$(W_{\mathbf{w}})^* (\mathbf{x}) = \sum_{i=1}^{n} w_{[i]} x_{(i)};$$

note that vectors $\mathbf{w}^*$ and $(x_{(1)}, \ldots, x_{(n)})$ are increasing, i.e., we multiply the smallest weight $w_{[1]}$ and the smallest input $x_{(1)}$, and so one, till the product of the greatest weight $w_{[n]}$ and the greatest input $x_{(n)}$;

- $(W_{\mathbf{w}})_* = \text{OWA}_{\mathbf{w}_*}$, where $\mathbf{w}_* = (w_{[n]}, \ldots, w_{[1]})$, and hence

$$(W_{\mathbf{w}})_* (\mathbf{x}) = \sum_{i=1}^{n} w_{[n-i+1]} x_{(i)};$$

here the greatest weight $w_{[n]}$ multiplies the smallest input $x_{(1)}$, etc.

- considering the input dependent symmetrization method $A_{B,K}$ introduced in the previous section for $B(x_1, x_2) = \frac{x_1 + 2x_2}{3}$ and $K = (id, rev)$, we have

$$(W_{\mathbf{w}})_{B,K} (\mathbf{x}) = \frac{1}{3} \left(\sum_{i=1}^{n} w_i x_{(i)} + 2 \sum_{i=1}^{n} w_i x_{(n-i+1)} \right) = \sum_{i=1}^{n} v_i x_{(i)} = \text{OWA}_{\mathbf{v}},$$

where, for $i = 1, \ldots, n$, the weight v_i is given by $v_i = \frac{1}{3}(w_i + 2w_{n-i+1})$.

For $n = 3$, let $\mathbf{w} = (0.5, 0.3, 0.2)$ and $\mathbf{x} = (0.4, 0.8, 0.6)$. Then

$$
\begin{aligned}
W_{\mathbf{w}}(\mathbf{x}) &= 0.56, \\
(W_{\mathbf{w}})_{\text{AM}}(\mathbf{x}) &= 0.6, \\
(W_{\mathbf{w}})^* (\mathbf{x}) &= \text{OWA}_{(0.2, 0.3, 0.5)}(0.4, 0.8, 0.6) = 0.66, \\
(W_{\mathbf{w}})_* (\mathbf{x}) &= \text{OWA}_{(0.5, 0.3, 0.2)}(0.4, 0.8, 0.6) = 0.54, \\
(W_{\mathbf{w}})_{B,K} (\mathbf{x}) &= \frac{1}{3} \left(W_{\mathbf{w}}(0.4, 0.6, 0.8) + 2W_{\mathbf{w}}(0.8, 0.6, 0.4) \right) \\
&= \text{OWA}_{(0.3, 0.3, 0.4)}(0.4, 0.8, 0.6) = 0.62.
\end{aligned}
$$

As another example, consider the weighted geometric mean

$$G_{(w_1, 1-w_1)}(x_1, x_2) = x_1^{w_1} x_2^{1-w_1},$$

which is not symmetric whenever $w_1 \neq \frac{1}{2}$. Note that now $n = n! = 2$. Let $B_p \in \mathcal{A}_2$ be a power-root operator given by

$$B_p(x_1, x_2) = \left(\frac{x_1^p + x_2^p}{2} \right)^{\frac{1}{p}},$$

where $p \in \mathbb{R}\setminus\{0\}$, and $B_0 = G$ is the geometric mean. Then, following Theorem 3, we have

$$\left(G_{(w_1, 1-w_1)}\right)_{B_p}(x_1, x_2) = \left(\frac{(x_1^{w_1} x_2^{1-w_1})^p + (x_1^{1-w_1} x_2^{w_1})^p}{2}\right)^{\frac{1}{p}}$$
$$= (x_1 x_2)^\alpha B_p(x_1^{1-2\alpha}, x_2^{1-2\alpha}),$$

where $\alpha = \min(w_1, 1 - w_1)$.

5   Concluding Remarks

We have introduced an axiomatic approach to symmetrization methods for aggregation functions. These methods belong either to input dependent methods (where the **x**-dependent permutation $(.) \in \mathcal{P}_n$ is considered) or to input independent methods. We have also shown two extremal symmetrization methods. Proposed approaches were illustrated by some examples, with a particular stress on the symmetrization of the weighted arithmetic means. Note that the extremal symmetrized weighted arithmetic means W^* and W_* can be seen as solutions of optimization methods and they can be related to the Hungarian algorithm [3] known from the area of linear optimization. For more details see [6]. Note also that one can further extend the problem of symmetrization of aggregation functions related to weighting vectors, where the symmetrization is related to both input vector **x** and the weighting vector **w**. This approach was initiated in [6] and we expect its deeper study in the near future.

Acknowledgments. The support of the grants APVV-14-0013 and VEGA 1/0682/16 is kindly announced.

References

1. Beliakov, G., Bustince Sola, H., Calvo Sánchez, T.: A Practical Guide to Averaging Functions. Springer, New York (2016)
2. Beliakov, G., Pradera, A., Calvo, T.: Aggregation Functions: A Guide for Practitioners. Springer, New York (2007)
3. Burkard, R.E., Dell'Amico, M., Martello, S.: Assignment Problems (Revised reprint). SIAM, Philadelphia (2012)
4. Calvo, T., Kolesárová, A., Komorníková, M., Mesiar, R.: Aggregation operators: properties, classes and construction methods. In: Aggregation Operators, Studies in Fuzziness and Soft Computing, vol. 97, pp. 3–104. Physica, Heidelberg (2002)
5. Grabisch, M., Marichal, J.L., Mesiar, R., Pap, E.: Aggregation Functions. Cambridge University Press, Cambridge (2009)
6. Mesiar, R., Stupňanová, A., Yager, R.R.: Extremal symmetrization of aggregation functions. Ann. Oper. Res. (2017). doi:10.1007/s10479-017-2471-x
7. Yager, R.R.: On ordered weighted averaging aggregation operators in multicriteria decisionmaking. IEEE Trans. Syst. Man Cybern. **18**(1), 183–190 (1988)

Event-Based Transformations of Capacities

Surajit Borkotokey[1(✉)], Radko Mesiar[2,3], and Jun Li[4]

[1] Department of Mathematics, Dibrugarh University,
Dibrugarh 786004, Assam, India
surajitbor@yahoo.com
[2] Faculty of Civil Engineering, Slovak University of Technology, Bratislava, Slovakia
mesiar@math.sk
[3] Faculty of Science, Palacký University, Olomouc, 12, 77146 Olomouc, Czechia
[4] School of Sciences, Communication University of China, Beijing 100024, China
lijun@cuc.edu.cn

Abstract. Event-based transformations of capacities are discussed. We study 4 particular event-based transformations of capacities and their convex closures. Due to commuting of convex combinations and our transformations, it is enough to examine four basic transformations of boolean capacities only.

Keywords: Boolean capacity · Capacity · Transformation · Fuzzy integral

1 Introduction

Capacities on a finite universe $\mathcal{N}$ of n criteria are commonly applied in multicriteria decision support [10–12]. Let us denote $\mathcal{N} = \{1, 2, ..., n\}$ without any loss of generality. By a capacity we mean a monotone set function $m : 2^{\mathcal{N}} \to [0, 1]$ satisfying two boundary conditions $m(\emptyset) = 0$ and $m(\mathcal{N}) = 1$. If, additionally, m is additive namely, $m(A \cup B) = m(A) + m(B)$ for any disjoint events $A, B \in 2^{\mathcal{N}}$, then m is a discrete probability. A mapping that maps a capacity onto another capacity determined by a fixed event is called an event-based capacity transformation. Recall that capacities in multicriteria decision support express the weights of groups of criteria, enabling to model the interactions between criteria. Subsequently, they are used to build utility functions by means of several kinds of fuzzy integrals, including the Choquet, Sugeno, concave, decomposition and other integrals [5,7,10,11,16]. Inspired by the idea of conditional probabilities, when, for any probability P and event B satisfying $P(B) > 0$, P is transformed into a new probability measure $P(./B)$, the concept of event-based transformation of capacities was proposed and studied in recent papers [2,8,14,15]. The aim of this contribution is a further development in this direction, studying four particular event-based transformations of capacities and their convex combinations, including links between these transformations. Our main results are presented in the Sects. 2 and 3 brings a complete description of the discussed event-based transformations in the case $n = 2$.

© Springer International Publishing AG 2017
V. Torra et al. (Eds.): MDAI 2017, LNAI 10571, pp. 33–39, 2017.
DOI: 10.1007/978-3-319-67422-3_4

2   Event Based Transforms

Let us denote by $\mathcal{M}_n$ the set of all capacities on $\mathcal{N}$. Considering the standard partial ordering of real set functions, $\mathcal{M}_n$ is a complete distributive lattice with the top element m^* and the bottom element m_*, given respectively by

$$m^*(E) = \begin{cases} 0 & \text{if } E = \emptyset \\ 1 & \text{otherwise} \end{cases} \tag{1}$$

and

$$m_*(E) = \begin{cases} 1 & \text{if } E = \mathcal{N} \\ 0 & \text{otherwise.} \end{cases} \tag{2}$$

Formally a mapping $\psi : \mathcal{M}_n \to \mathcal{M}_n$ is called a capacity transformation. If for any event $B \subseteq \mathcal{N}$, we have a system $\boldsymbol{\Phi} = (\psi_B)_{B \subseteq \mathcal{N}}$ of capacity transformations, where ψ_B depends on B only, then $\boldsymbol{\Phi}$ is called an event-based capacity transformation system, and ψ_B is called a B-based capacity transform.

Note that the conditional probability transformation is not a probability transformation, in general. This is caused by the failure of this approach in the case when $P(B) = 0$. On the other hand, the formula $P(B \cap E)/P(B \cup E^c)$ (with convention $0/0 = 1$) can be applied for any probability measure P and any events $B, E \subseteq \mathcal{N}$, preserving the monotonicity and boundary conditions in E (i.e., defining a capacity), but violating the additivity, in general. This approach can be applied to any capacity $m \in \mathcal{M}_n$, thus yielding a capacity transformation τ given by, for any fixed $B \subseteq \mathcal{N}$, $\tau_B(m)(E) = m(B \cap E)/m(B \cup E^c)$. Recall now the valuation property (i.e., modularity) of capacities characterized by the validity of equation (for any subsets B and E of $\mathcal{N}$)

$$m(E) + m(B) = m(B \cup E) + m(B \cap E),$$

or, equivalently

$$m(E) = m(B \cup E) - m(B) + m(B \cap E). \tag{3}$$

Interestingly, considering the right-hand side of (3) for a fixed B, and an arbitrary capacity $m \in \mathcal{M}_n$, the set function $\epsilon_B(m)$ defined on $2^{\mathcal{N}}$ by

$$\epsilon_B(m)(E) = m(B \cup E) - m(B) + m(B \cap E)$$

is a capacity transformation. We consider this capacity transformation and some related capacity transformations in the next example.

Example 1. Fix $B \in \mathcal{N}$. Denote by ϵ_B, α_B, β_B and γ_B the maps over $\mathcal{M}_n$ defined by

$$\epsilon_B(m)(E) = m(B \cup E) - m(B) + m(B \cap E) \tag{4}$$

$$\alpha_B(m)(E) = m(B \cup E) - m(B \cap E^c) \tag{5}$$

$$\beta_B(m)(E) = 1 - m(B \cup E^c) + m(B \cap E) \tag{6}$$

$$\gamma_B(m)(E) = 1 - m(B \cup E^c) + m(B) - m(B \cap E^c) \tag{7}$$

for $m \in \mathcal{M}_n$ and $E \subseteq \mathcal{N}$. Then each of ϵ_B, α_B, β_B and γ_B is a B-based capacity transformation.

Observe that each of these transformations is related to some representation of E, namely,

$$\epsilon_B :\rightarrow E = ((B \cup E) \setminus B) \cup (B \cap E); \qquad \alpha_B :\rightarrow E = (B \cup E) \setminus (B \cap E^c)$$
$$\beta_B :\rightarrow E = (\mathcal{N} \setminus (B \cup E^c)) \cup (B \cap E); \qquad \gamma_B :\rightarrow E = (\mathcal{N} \setminus (B \cup E^c)) \cup (B \setminus (B \cap E^c)).$$

Moreover for every $(a_1, a_2, a_3, a_4) \in [0,1]^4$ with $\sum_{i=1}^4 a_i = 1$, $a_1 \epsilon_B + a_2 \alpha_B + a_3 \beta_B + a_4 \gamma_B$ is also a B-based capacity transformation.

Remark 1.

(i) Observe that if $B = \mathcal{N}$ then $\epsilon_{\mathcal{N}}(m) = \beta_{\mathcal{N}}(m) = m$ and $\alpha_{\mathcal{N}}(m) = \gamma_{\mathcal{N}}(m) = m^d$. Similarly for $B = \emptyset$, $\epsilon_{\emptyset}(m) = \alpha_{\emptyset}(m) = m$, $\beta_{\emptyset}(m) = \gamma_{\emptyset}(m) = m^d$, where m^d denotes the dual measure given by $m^d(E) = 1 - m(E^c)$.

(ii) We focus on capacity transformations introduced in Example 1 due to the fact that they are compatible with the convex structure of the set $\mathcal{M}_n$ of all capacities on $\mathcal{N}$. Note that this is not the case of many other possible capacity transformations, such as the τ transformation introduced above.

Observe that the transformation β_B is based on the Lukasiewicz implication [1,4,6,9], while the transformation α_B can be derived by means of the Lukasiewicz coimplication [3]. Note also that while the transformation β_B was introduced and studied already in [2,7,13,14] the remaining transformations ϵ_B, α_B and γ_B are, up to our best knowledge, presented for the first time in this paper. It is not difficult to see that, given $B \subseteq \mathcal{N}$, the transformations $\epsilon_B, \alpha_B, \beta_B$ and γ_B are independent in that sense that none of them can be obtained as a convex combination of the remaining three transformations. Moreover, any composition of two of these transformations results into a B-based transformation. The structure of these compositions is described in the next Theorem.

Theorem 1. *For all B, the set $G_B = \{\epsilon_B, \alpha_B, \beta_B, \gamma_B\}$ is an Abelian group with neutral element ϵ_B under the composition "$\circ$" operation of functions, given by the following table.*

$\circ$	ϵ_B	α_B	β_B	γ_B
ϵ_B	ϵ_B	α_B	β_B	γ_B
α_B	α_B	ϵ_B	γ_B	β_B
β_B	β_B	γ_B	ϵ_B	α_B
γ_B	γ_B	β_B	α_B	ϵ_B

Composition Table

Moreover, $G_B \cong \mathbb{Z}_2^+ \times \mathbb{Z}_2^+$ so that $\epsilon_B \sim (0,0)$, $\alpha_B \sim (1,0)$, $\beta_B \sim (0,1)$ and $\gamma_B \sim (1,1)$.

The convex closure of G_B denoted by H_B is given by

$$H_B = \left\{ a_1 \epsilon_B + a_2 \alpha_B + a_3 \beta_B + a_4 \gamma_B \mid (a_1, a_2, a_3, a_4) \in [0,1]^4, \ \sum_{i=1}^4 a_i = 1 \right\} \quad (8)$$

It follows from the definition of H_B that each of its elements which is of the form $a_1\epsilon_B + a_2\alpha_B + a_3\beta_B + a_4\gamma_B$ can be identified with the vector $\mathbf{a} = (a_1, a_2, a_3, a_4) \in [0,1]^4$, $a_1 + a_2 + a_3 + a_4 = 1$. Thus under this new notation, we have $\epsilon_B \sim (1,0,0,0)$ and $\frac{\epsilon_B + \alpha_B + \beta_B + \gamma_B}{4} \sim (\frac{1}{4}, \frac{1}{4}, \frac{1}{4}, \frac{1}{4})$. The composition $\circ$ of transformations $\phi, \psi \in H_B$, $\eta = \phi \circ \psi$, is defined, when the vector representation $\phi \sim \mathbf{a}$, $\psi \sim \mathbf{b}$ and $\eta \sim \mathbf{c}$ is considered, is defined as follows. For $\mathbf{a}, \mathbf{b} \in [0,1]^4$, $\mathbf{a} \circ \mathbf{b} = \mathbf{b} \circ \mathbf{a} = \mathbf{c} = (c_1, c_2, c_3, c_4)$, where $c_1 = \sum_{i=1}^{4} a_i b_i$ and $c_k = \sum_{\substack{i \neq j \\ |i+j-5|=4-k}} a_i b_j$, $k = 2,3,4$. The next result follows immediately.

Theorem 2. H_B is an Abelian semigroup with neutral element ϵ_B and annihilator (zero element) $\frac{\epsilon_B + \alpha_B + \beta_B + \gamma_B}{4}$.

Theorem 3. Let $\mathcal{B}_n = \{m \in \mathcal{M}_n | \; Range(m) = \{0,1\}\}$ be the set of all boolean capacities. For all $\phi \in G_B$, $B \subseteq \mathcal{N}$ $B \subseteq \mathcal{N}$, the transformation ϕ can be restricted to act on boolean capacities from $\mathcal{B}_n$, $\phi|_{\mathcal{B}_n} : \mathcal{B}_n \to \mathcal{B}_n$ (i.e., ϕ maps boolean capacities onto boolean capacities).

Proof. The proof follows immediately from the definitions of the transformations.

Theorem 4. For all $\phi \in H_B$, $B \subseteq \mathcal{N}$, ϕ commutes with all convex combinations,

$$\phi(\sum_{i=1}^{k} \lambda_i m_i) = \sum_{i=1}^{k} \lambda_i \phi(m_i)$$

where $\lambda_i \in [0,1]$ for each $i \in \mathcal{N}$ and $\sum_{i=1}^{n} \lambda_i = 1$.

Recall that Dirac measures $\delta_1, ... \delta_n : 2^{\mathcal{N}} \to [0,1]$ are given, for $i \in \mathcal{N}$, by

$$\delta_i(A) = \begin{cases} 1 & \text{if } i \in A \\ 0 & \text{otherwise.} \end{cases} \tag{9}$$

Directly one can check that for any δ_i, $B \subseteq \mathcal{N}$, $\phi \in G_B$ we have $\phi(\delta_i) = \delta_i$. It follows that $\forall \phi \in H_B$, $\phi(\delta_i) = \delta_i$. Moreover, it has been shown that only Dirac measures are totally invariant (i.e., for any $B \subseteq \mathcal{N}$, $\phi \in H_B$, and $m \in \mathcal{M}_n$: $\phi(m) \equiv m$). From the fact that $\mathcal{M}_n$ is the convex closure of $\mathcal{B}_n$ (see, for example, [12]) and due to Theorem 4, it is enough to know, for fixed $B \subseteq \mathcal{N}$, transforms $\psi(\mu)$ for all boolean capacities $\mu \in \mathcal{B}_n$ and $\psi \in G_B$, then for any $\eta \sim \mathbf{a} = (a_1, a_2, a_3, a_4) \in H_B$ and $m \in \mathcal{M}_n$, $m = \sum_{i=1}^{k} \lambda_i \mu_i$, $\mu_i \in \mathcal{B}_n$, we have

$$\eta(m) = a_1 \sum_{i=1}^{k} \lambda_i \epsilon_B(\mu_i) + a_2 \sum_{i=1}^{k} \lambda_i \alpha_B(\mu_i) + a_3 \sum_{i=1}^{k} \lambda_i \beta_B(\mu_i) + a_4 \sum_{i=1}^{k} \lambda_i \gamma_B(\mu_i)$$

$$\tag{10}$$

3   Complete Description of Tranformations of Capacities from $\mathcal{M}_2$

In this section, we describe completely the action of transformations from H_B in the case when $n = 2$, considering arbitrary $B \subseteq \{1,2\}$ and arbitrary capacity $m \in \mathcal{M}_2$.

Example 2. For $n = 2$, we have only four boolean measures : δ_1, δ_2, m^* and m_*. δ_1 and δ_2 are totally invariant; each $m \in \mathcal{M}_2$ is uniquely determined by $(m(\{1\}); m(\{2\}))$. Thus we will use, as an abbreviation, the notation $m \sim (m(\{1\}); m(\{2\}))$. Observe that due to this representation there is a one-to-one correspondence between the capacities from $\mathcal{M}_2$ and couples from $[0,1]^2$. Due to the total invariantness of Dirac functions, we need to know the B-transforms of extremal capacities m^* and m_* only. We will describe these transformations for any $B \subseteq \{1,2\}$ and all basic transformations $\epsilon_B, \alpha_B, \beta_B$ and γ_B. For $m^* \sim (1,1)$, it holds:

$\epsilon_\emptyset(m^*) = \epsilon_{\{1,2\}}(m^*) = m^*; \; \epsilon_{\{1\}}(m^*) \sim (1,0) \sim \delta_1; \; \epsilon_{\{2\}}(m^*) \sim (0,1) \sim \delta_2.$
$\alpha_\emptyset(m^*) = m^* \sim (1,1); \; \alpha_{\{1,2\}}(m^*) = m_* \sim (0,0), \; \alpha_{\{1\}}(m^*) \sim (1,0) \sim \delta_1;$
$\alpha_{\{2\}}(m^*) \sim (0,1) \sim \delta_2.$
$\beta_\emptyset(m^*) = m_* \sim (0,0); \; \beta_{\{1,2\}}(m^*) = m^* \sim (1,1), \; \beta_{\{1\}}(m^*) \sim (1,0) \sim \delta_1;$
$\beta_{\{2\}}(m^*) \sim (0,1) \sim \delta_2.$
$\gamma_\emptyset(m^*) = m_* = \gamma_{\{1,2\}}(m^*) \sim (0,0), \; \gamma_{\{1\}}(m^*) \sim (1,0) \sim \delta_1; \; \gamma_{\{2\}}(m^*) \sim (0,1) \sim \delta_2.$

Summarising, for any $\phi \in H_B$, $\phi \sim \mathbf{a} = (a_1, a_2, a_3, a_4)$ it holds : if $B = \emptyset$ then $\phi(m^*) = s_{a_1+a_2} \sim (a_1 + a_2, a_1 + a_2)$, where $s_c \sim (c,c)$ is a symmetric capacity on $\{1,2\}$ such that $s_c(\{1\}) = s_c(\{2\}) = c$; when $B = \{1,2\}$ we have $\phi(m^*) = s_{a_1+a_3} \sim (a_1+a_3, a_1 \mid a_3)$; when $B = \{1\}$ we have $\phi(m^*) = \delta_1$ and when $B = \{2\}$ we have $\phi(m^*) = \delta_2$; similarly, for m_* it holds: when $B = \emptyset, \phi(m_*) = s_{a_3+a_4}$; when $B = \{1,2\}$, it holds $\phi(m_*) = s_{a_2+a_4}$; when $B = \{1\}$ we have $\phi(m_*) = \delta_2$ and finally when $B = \{2\}$ we have $\phi(m_*) = \delta_1$.

Now we know $\phi(\mu)$ for all $\mu \in \mathcal{B}_2$. For any capacity $m \in \mathcal{B}_2$, $m \sim (a,b)$, we have the following cases.

Case I:

If $a = b$, i.e., $m = s_a$, then $m = am^* + (1-a)m_*$ and we have, when $B = \emptyset$, it holds $\phi(m) = a\phi(m^*) + (1-a)\phi(m_*) = as_{a_1+a_2} + (1-a)s_{a_3+a_4} = s_{a(a_1+a_2)+(1-a)(a_3+a_4)}$; when $B = \{1,2\}$ we have $\phi(m) = s_{a(a_1+a_3)+(1-a)(a_2+a_4)}$; when $B = \{1\}$ we have $\phi(m) = a\delta_1 + (1-a)\delta_2 \sim (a, 1-a)$ and finally when $B = \{2\}$ we have $\phi(m) = a\delta_2 + (1-a)\delta_1 \sim (1-a, a)$.

Case II:

If $a > b$, then $m = bm^* + (a-b)\delta_1 + (1-a)m_*$ and we have, when $B = \emptyset$, it holds: $\phi(m) = b\phi(m^*) + (a-b)\phi(\delta_1) = bs_{a_1+a_2} + (a-b)\delta_1 + (1-a) + (1-a)\phi(m_*) \sim (b(a_1+a_2)+a-b)+(1-a)(a_3+a_4), b(a_1+a_2))+(1-a)(a_3+a_4)$; when $B = \{1,2\}$ we have $\phi(m) \sim (b(a_1+a_3)+a-b)+(1-a)(a_2+a_4), b(a_1+a_3)+(1-a)(a_2+a_4))$; when $B = \{1\}$ we have $\phi(m) = a\delta_1 + (1-a)\delta_2 \sim (a, 1-a)$ and finally when $B = \{2\}$ we have $\phi(m) = b\delta_2 + (a-b)\delta_1 + (1-a)\delta_1 \sim (1-b, b)$.

Case III:

If $a < b$, then $m = am^* + (b-a)\delta_2 + (1-b)m_*$ and we have, when $B = \emptyset$, it holds : $\phi(m) \sim (a(a_1+a_2)+(1-b)(a_3+a_4), a(a_1+a_2)+b-a+(1-b)(a_3+a_4)$; when $B = \{1,2\}$ we have $\phi(m) \sim (a(a_1+a_2)+(1-b)(a_2+a_4), a(a_1+a_3)+b-a+(1-b)(a_2+a_4))$; when $B = \{1\}$, it holds $\phi(m) \sim (a, 1-a)$ and finally when

$B = \{2\}$ we have $\phi(m) = \sim (1 - b, b)$. Observe that always if $a + b = 1$, i.e., if m is additive, we have $\phi(m) = m$, independently of $B \subseteq \{1, 2\}$.

4   Concluding Remarks

We have introduced four types of event-based transformations of capacities and studied their relationships, including the structure of their convex closure. While our four transformations form an Abelian group $G_B \cong \mathbb{Z}_2^+ \times \mathbb{Z}_2^+$, the corresponding convex closure H_B is an Abelian semigroup with an annihilator. We have examplified all discussed transformations in the simpliest non-trivial case, i.e., when $n = 2$. Recall that the study of event-based transformations of capacities was inspired by conditional probabilities. Having in mind importance and applicability of Bayesian approaches based on conditional probabilities, we believe that also the idea of event-based transformations of capacities will grow and bring new views into the large area of multicriteria decision support.

Acknowledgement. This work was partially funded by the grant APVV-14-0013 and VEGA 1/0420/15, the National Natural Science Foundation of China (grants No. 11371332 and No. 11571106) and SAIA, Slovakia.

References

1. Baczyński, M., Jayaram, B.: Fuzzy Implications, Studies in Fuzziness and Soft Computing. Springer, Berlin (2008)
2. Borkotokey, S., Komorníková, M., Li, J., Mesiar, R.: Aggregation functions, similarity and fuzzy measures. In: AGOP-2017, Advances in intelligent systems and computing, Springer, Cham (2017, to appear)
3. De Baets, B.: Coimplications, the forgotten connectives. Tatra Mt. Mat. Publ. **12**, 229–240 (1997)
4. De Baets, B., Mesiar, R.: Residual implicators of continuous t-norms. In: Proceedings of EUFIT 1996, pp. 27–31. Aachen (1996)
5. Radojević, D.: Logical interpretation of discrete Choquet integral defined by general measures. Int. J. Uncertain. Fuzziness Knowl. Based Syst. **7**(6), 577–588 (1999)
6. Durante, F., Klement, E.P., Mesiar, R., Sempi, C.: Conjunctors and their residual implicators: characterizations and construction methods. Mediterr. J. Math. **4**, 343–356 (2007)
7. Even, Y., Lehrer, E.: Decomposition-integral, unifying Choquet and Concave integrals. Econ. Theor. **56**, 33–58 (2014)
8. Kouchakinejad, F., Šipošová, A.: On some transformations of fuzzy measures. Tatra Mt. Mat. Publ. (2017, to appear)
9. Gottwald, S.: A Treatise on Many-Valued Logics. Studies in Logic and Computation, vol. 9. Research Studies Press, Baldock (2001)
10. Grabisch, M.: Set Functions, Games and Capacities in Decision Making, Theory and Decision Library, vol. 46. Springer, Berlin (2016)
11. Grabisch, M., Labreuche, C.: A decade of application of the Choquet and Sugeno integrals in multi-criteria decision aid. Ann. Oper. Res. **175**, 247–286 (2010)

12. Greco, S., Ehrgott, M., Figueira, J.R. (eds.): Multiple Criteria decision Analysis: State of the Art Surveys. Springer, New York (2005)
13. Klement, E.P., Mesiar, R., Pap, E.: Triangular Norms. Kluwer, Dordrecht (2000)
14. Jin, L.S., Mesiar, R., Borkotokey, S., Kalina, M.: Certainty aggregation and the certainty fuzzy measures. IEEE Trans. Fuzzy Syst. (2017, submitted)
15. Mesiar, R., Borkotokey, S., Jin, L.S., Kalina, M.: Aggregation functions and capacities. Fuzzy Sets Syst. (2017). http://dx.doi.org/10.1016/j.fss.2017.08.007
16. Mesiar, R., Stupňanová, A.: Decomposition integrals. Int. J. Approx. Reason. **54**, 1252–1259 (2013)

Pan-Integrals Based on Optimal Measures

Jun Li[1(✉)], Yao Ouyang[2], and Minhao Yu[1]

[1] School of Sciences, Communication University of China, Beijing 100024, China
{lijun,yuminhao}@cuc.edu.cn
[2] Faculty of Science, Huzhou University, Huzhou 313000, Zhejiang, China
oyy@zjhu.edu.cn

Abstract. The pan-integrals are based on a special type of commutative isotonic semiring $(\overline{R}_+, \oplus, \otimes)$ and the monotone measures μ defined on a measurable space $(X, \mathcal{A})$. On the other hand, based on a pan-addition $\oplus$ each monotone measure μ generates a new monotone measure $\mu_\oplus$ which is called the $\oplus$-optimal measure (to μ and $\oplus$). Such monotone measure $\mu_\oplus$ is greater than or equal to μ and it is super-$\oplus$-additive (i.e., $\mu_\oplus(A \cup B) \geq \mu_\oplus(A) \oplus \mu_\oplus(B)$ whenever $A, B \in \mathcal{A}$, $A \cap B = \emptyset$). In this note, we shall present some new properties of the pan-integral. It is shown that the pan-integral with respect to μ coincides with the pan-integral with respect to $\mu_\oplus$ on a given pan-space $(X, \mathcal{A}, \mu, \overline{R}_+, \oplus, \otimes)$. As a special case of this result, we show that the $\oplus$-optimal measure derived from μ is totally balanced for the pan-integrals.

Keywords: Motonone measure · Pan-integral · Optimal measure · Pan-addition · Pan-multiplication · Super-$\oplus$-additivity

1 Introduction

In nonlinear integral theory there are several prominent integrals, for example, the Choquet integral [3], the Sugeno integral [19], the pan-integral [23] and the concave integral introduced by Lehrer [8], etc. The pan-integral introduced in [23] (see also [21]) is based on a monotone measure μ and relates to a commutative isotonic semiring $(\overline{R}_+, \oplus, \otimes)$, where $\oplus$ is a pan-addition and $\otimes$ is a pan-multiplication related by the distributivity property [21,23]. This integral generalizes the Lebesgue integral and Sugeno integral, i.e., when considering a σ-additive measure m and the commutative isotonic semiring $(\overline{R}_+, +, \cdot)$, the Lebesgue integral coincides the pan-integral based on the usual addition $+$ and multiplication $\cdot$, and considering a monotone measure μ and the commutative isotonic semiring $(\overline{R}_+, \vee, \wedge)$, the Sugeno integral coincides the pan-integral with respect to $(\vee, \wedge)$. The researches on this topic can be also found in [1,5,12,16,17,20,24].

In this paper we present some new properties of the pan-integrals. This study comes from the following problem related to the Lebesgue integral and the concave integral: Let $(X, m, \mathcal{A})$ be a measure space. For any $A \in \mathcal{A}$, considering Lebesgue integration of the characteristic function χ_A, then we have

© Springer International Publishing AG 2017
V. Torra et al. (Eds.): MDAI 2017, LNAI 10571, pp. 40–50, 2017.
DOI: 10.1007/978-3-319-67422-3_5

$\int \chi_A dm = m(A)$. Note that when we consider nonlinear integrals with respect to monotone measure μ, integrating the characteristic function χ_A the above discussed equality holds also for several nonlinear integrals with respect to monotone measure, including the Choquet integral, Shilkret integral [18] or Sugeno integral [19] (assuming $\mu(X) = 1$), and all integrals which are both universal [7] and decomposable [15] (see also [6]). In general, for the concave integrals or the pan-integrals, we have always integral from characteristic function greater or equal the measure of the corresponding measurable set, and thus the violation of the equality, which is possible for both types of integrals, necessarily means that integral is greater than the measure of the corresponding set. If a monotone measure ν satisfies $\int \chi_A d\nu = \nu(A)$ for every $A \in \mathcal{A}$, then ν is called *totally balanced* (with respect to the involved integral), see [9]. In the case of the concave integral, Lehrer study the above mentioned problem [9].

Given a monotone measure ν over $(X, \mathcal{A})$, one can define a monotone measure $\hat{\nu}_{cav}$ generated by ν, by using the concave integral [9], as follows:

$$\hat{\nu}_{cav}(A) = (cav) \int \chi_A d\nu, \quad \forall A \in \mathcal{A}.$$

Then the following properties hold [9]:

(i) $\nu \leq \hat{\nu}_{cav}$, i.e., $\forall A \in \mathcal{A}$, $\nu(A) \leq \hat{\nu}_{cav}(A)$;
(ii) for all nonnegative measurable function f,

$$(cav) \int f d\hat{\nu} = (cav) \int f d\nu;$$

(iii) $\hat{\nu}_{cav}(A) = \int^{cav} \chi_A d\hat{\nu}_{cav}$, i.e., $\hat{\hat{\nu}}_{cav} = \hat{\nu}_{cav}$.

This show that the monotone measure $\hat{\nu}_{cav}$ is totally balanced (with respect to the concave integral).

In this paper we shall make some researches on this topic in the case of pan-integrals.

Let $(X, \mathcal{A}, \mu, \overline{R}_+, \oplus, \otimes)$ be a given pan-space [21]. Similarly, if we define a new set function $\hat{\mu}_{pan}$ generated by μ, as follows:

$$\hat{\mu}_{pan}(A) = (p) \int \chi_A d\mu, \quad \forall A \in \mathcal{A},$$

then $\hat{\mu}_{pan}$ is a monotone measure, and it is just the $\oplus$-optimal measure $\mu_\oplus$ to μ (see [10]), i.e.,

$$\mu_\oplus(A) = (p) \int \chi_A d\mu, \quad \forall A \in \mathcal{A}.$$

We shall prove that the pan-integral with respect to μ coincides with the pan-integral with respect to $\mu_\oplus$. Thus we only need to discuss the pan-integrals with respect to $\mu_\oplus$. As a special case of the result, we can see that the $\oplus$-optimal measure derived from μ is totally balanced for the pan-integrals.

2   Preliminaries

Let X be a nonempty set, $\mathcal{A}$ a σ-algebra of subsets of X, $R_+ = [0, +\infty)$ and $\overline{R}_+ = [0, +\infty]$. A set function $\mu : \mathcal{A} \to \overline{R}_+$ is called a *monotone measure* on $(X, \mathcal{A})$ [21], if it satisfies the following conditions:

(1) $\mu(\emptyset) = 0$ and $\mu(X) > 0$;
(2) $\mu(A) \le \mu(B)$ whenever $A \subset B$ and $A, B \in \mathcal{A}$.

The triple $(X, \mathcal{A}, \mu)$ is called a monotone measure space [17, 21].

In this paper we restrict our discussion on a fixed measurable space $(X, \mathcal{A})$. Unless stated otherwise all the subsets mentioned are supposed to belong to $\mathcal{A}$. $\mathcal{M}$ denotes the set of all monotone measures defined on $(X, \mathcal{A})$.

The concept of a pan-integral involves two binary operations, the pan-addition $\oplus$ and pan-multiplication $\otimes$ on non-negative real numbers [21, 23]

Definition 1. An binary operation $\oplus$ on $\overline{R}_+$ is called a pan-addition if it satisfies the following requirements:

(PA1) $a \oplus b = b \oplus a = a$ (commutativity);
(PA2) $(a \oplus b) \oplus c = a \oplus (b \oplus c)$ (associativity);
(PA3) $a \le c$ and $b \le d$ imply that $a \oplus b \le c \oplus d$ (monotonicity);
(PA4) $a \oplus 0 = a$ (neutral element);
(PA5) $a_n \to a$ and $b_n \to b$ imply that $a_n \oplus b_n \to a \oplus b$ (continuity).

Definition 2. Let $\oplus$ be a given pan-addition on $\overline{R}_+$. A binary operation $\otimes$ on $\overline{R}_+$ is said to be a pan-multiplication corresponding to $\oplus$ if it satisfies the following properties:

(PM1) $a \otimes b = b \otimes a$ (commutativity);
(PM2) $(a \otimes b) \otimes c = a \otimes (b \otimes c)$ (associativity);
(PM3) $a \otimes (b \oplus c) = (a \otimes b) \oplus (a \otimes c)$ (distributive law);
(PM4) $a \le b$ implies $(a \otimes c) \le (b \otimes c)$ for any c (monotonicity);
(PM5) $a \otimes b = 0 \Leftrightarrow a = 0$ or $b = 0$ (annihilator);
(PM6) there exists $e \in \overline{R}_+$ such that $e \otimes a = a$ for any $a \in \overline{R}_+$ (neutral element);
(PM7) $a_n \to a \in [0, \infty)$ and $b_n \to b \in [0, \infty)$ imply $(a_n \otimes b_n) \to (a \otimes b)$ (continuity).

When $\oplus$ is a pseudo-addition on $\overline{R}_+$ and $\otimes$ is a pseudo-multiplication (with respect to $\oplus$) on $\overline{R}_+$, the triple $(\overline{R}_+, \oplus, \otimes)$ is called a *commutative isotonic semiring* (with respect to $\oplus$ and $\otimes$) and $(X, \mathcal{A}, \mu, \overline{R}_+, \oplus, \otimes)$ is called a *pan-space* [21].

Notice that similar operations called pseudo-addition and pseudo-multiplication can be found in the literature [1, 2, 4, 5, 11, 14, 17, 20, 24].

Let $\oplus$ be a given pan-addition, $\mu \in \mathcal{M}$. μ is called

(1) *sub-$\oplus$-additive*, if $\mu(A \cup B) \le \mu(A) \oplus \mu(B)$ whenever $A, B \in \mathcal{A}$;
(2) *super-$\oplus$-additive*, if $\mu(A \cup B) \ge \mu(A) \oplus \mu(B)$ whenever $A, B \in \mathcal{A}$, $A \cap B = \emptyset$;
(3) *$\oplus$-additive*, if $\mu(A \cup B) = \mu(A) \oplus \mu(B)$ for any $A, B \in \mathcal{A}$, $A \cap B = \emptyset$.

3   Optimal Measures

We recall the concept of optimal measure with respect to a monotone measure μ and a pan-addition $\oplus$, see [10].

Definition 3. Given a pan-addition $\oplus$ and $\mu \in \mathcal{M}$. The set function $\mu_\oplus : \mathcal{A} \to \overline{R}_+$ defined by

$$\mu_\oplus(A) = \sup\left\{\bigoplus_{i=1}^{n}\mu(A \cap E_i) \mid \{E_i\}_{i=1}^{n} \in \hat{\mathcal{P}}\right\}, \quad A \in \mathcal{A}. \tag{1}$$

is called the $\oplus$-optimal measure to μ and $\oplus$. Moreover, μ is called a $\oplus$-optimal measure whenever $\mu = \mu_\oplus$.

The following properties (i)–(iv) are due to [10].

(i) For any $\mu \in \mathcal{M}$, then $\mu_\oplus \in \mathcal{M}$.
(ii) For each $A \in \mathcal{A}$,

$$\mu_\oplus(A) = \sup\left\{\bigoplus_{i=1}^{n}\mu(F_i) \mid \{F_i\}_{i=1}^{n} \in A \cap \hat{\mathcal{P}}\right\},$$

where $A \cap \hat{\mathcal{P}}$ denotes the set of all finite measurable partitions of A.
(iii) $\mu \leq \mu_\oplus$, i.e., $\mu(A) \leq \mu_\oplus(A)$ for each $A \in \mathcal{A}$.
(iv) $(\mu_\oplus)_\oplus = \mu_\oplus$ for each $\mu \in \mathcal{M}$.

Proposition 1. *Given a pan-addition $\oplus$ and $\mu \in \mathcal{M}$. Then μ is $\oplus$-optimal measure if and only if μ is super-$\oplus$-additive.*

From the above Proposition 1 and property (iv), $\mu_\oplus$ is $\oplus$-optimal measure, and hence $\mu_\oplus$ is super-$\oplus$-additive.

Proposition 2. *Given a pan-addition $\oplus$ and $\mu \in \mathcal{M}$. Then $\mu_\oplus$ is $\oplus$-optimal measure, i.e., $(\mu_\oplus)_\oplus = \mu_\oplus$, and hence $\mu_\oplus$ is super-$\oplus$-additive.*

Proof. It follows from Proposition 1 and property (iv).

4   Pan-Integrals

The concept of *pan-integral* based on pan-addition $\oplus$ and pan-multiplication $\otimes$ was introduced in [23] (see also [21, 22, 24]).

A finite measurable partition of X is a finite disjoint system of sets $\{A_i\}_{i=1}^{n} \subset \mathcal{A}$ such that $A_i \cap A_j = \emptyset$ for $i \neq j$ and $\cup_{i=1}^{n}A_i = X$. The set of all finite measurable partitions of X is denoted by $\hat{\mathcal{P}}$.

We restrict the discussion to a given pan-space $(X, \mathcal{A}, \mu, \overline{R}_+, \oplus, \otimes)$.

Let $A \in \mathcal{A}$. A (generalized) real-valued function defined on X given by

$$\chi_A = \begin{cases} e & \text{if } x \in A, \\ 0 & \text{otherwise} \end{cases}$$

is called the *pseudo-characteristic function of A*, where e is fixed neutral element of $\otimes$. A real-valued function defined on X given by

$$s(x) = \bigoplus_{i=1}^{n} \left[a_i \otimes \chi_{A_i}(x) \right]$$

is called a *pseudo-simple function* (with respect to $(\oplus, \otimes)$), where $a_i \in R_+, i = 1, 2, \ldots, n$, $\{A_i\}_{i=1}^{n} \in \hat{\mathcal{P}}$.

Let $\mathcal{S}$ denote the set of all pseudo-simple functions on X, and $\mathcal{F}^+$ denote the class of all real-valued nonnegative measurable functions on $(X, \mathcal{A})$. Obviously, for any $A \in \mathcal{A}$, $\chi_A \in \mathcal{S}$ and $\mathcal{S} \subset \mathcal{F}^+$.

Definition 4. Let $(X, \mathcal{A}, \mu, \overline{R}_+, \oplus, \otimes)$ be a pan-space, $A \in \mathcal{A}$ and $f \in \mathcal{F}^+$. The pan-integral of f on A with respect to μ, is defined by

$$(p) \int_A^{(\oplus, \otimes)} f d\mu = \sup_{\mathcal{E} \in \hat{\mathcal{P}}} \left\{ \bigoplus_{E \in \mathcal{E}} \left[\left(\inf_{x \in A \cap E} f(x) \right) \otimes \mu(A \cap E) \right] \right\}. \tag{2}$$

When $A = X$, $(p) \int_A^{(\oplus, \otimes)} f d\mu$ is written as $(p) \int^{(\oplus, \otimes)} f d\mu$.

If $(p) \int_A^{(\oplus, \otimes)} f d\mu < \infty$, then we say that f is pan-integrable on A. When f is pan-integrable on X, we simply say that f is pan-integrable.

The above pan-integral of f can be expressed as the following form:

$$(p) \int_A^{(\oplus, \otimes)} f d\mu = \sup \left\{ \bigoplus_{i=1}^{n} (\lambda_i \otimes \mu(A_i)) : \bigoplus_{i=1}^{n} [\lambda_i \otimes \chi_{A_i}] \leq f, \{A_i\}_{i=1}^{n} \in \hat{\mathcal{P}} \right\}.$$

Note that in the case of commutative isotonic semiring $(\overline{R}_+, \vee, \wedge)$, Sugeno integral [19] is recovered, while for $(\overline{R}_+, \vee, \cdot)$, Shilkret integral [18] is covered by the pan-integral in Definition 4.

In the rest of the paper we assume that the neutral element e in the commutative isotonic semiring $(\overline{R}_+, \oplus, \otimes)$ is finite, i.e., $e < +\infty$. In the case of $e = +\infty$, $\oplus = \vee$ (see [14]).

The following recall some basic properties of the pan-integrals [16, 21, 24].

Proposition 3. *Let $f, g \in \mathcal{F}_+, A \in \mathcal{A}$ and $a \in R_+$. Then we have the following:*

(1) $(p) \int_A^{(\oplus, \otimes)} f d\mu = (p) \int^{(\oplus, \otimes)} f \otimes \chi_A d\mu$;

(2) if $f \leq g$, then $(p) \int_A^{(\oplus, \otimes)} f d\mu \leq (p) \int_A^{(\oplus, \otimes)} g d\mu$;

(3) $(p) \int^{(\oplus, \otimes)} a \otimes f d\mu \geq a \otimes (p) \int^{(\oplus, \otimes)} f d\mu$, *in particular, for $A \in \mathcal{A}$;*
    $(p) \int^{(\oplus, \otimes)} a \otimes \chi_A d\mu \geq a \otimes (p) \int^{(\oplus, \otimes)} \chi_A d\mu$;

(4) $(p) \int^{(\oplus, \otimes)} \chi_A d\mu \geq \mu(A)$.

Proposition 4. *Let $\mu \in \mathcal{A}$ and $\mu_\oplus$ be the $\oplus$-optimal measure to μ. Then*

$$\mu_\oplus(A) = (p) \int^{(\oplus,\otimes)} \chi_A d\mu.$$

Proof. It follows from definitions of $\mu_\oplus$ and the pan-integral, see also [10]. $\blacksquare$

5   Pan-Integral Based on Optimal Measures

Given a monotone measure $\mu \in \mathcal{M}$, similar to the discussion of the concave integrals [9], we study the totally balance of a monotone measure $\hat{\mu}_{pan}$ generated by μ and the pan-integral. As we have mentioned in Sect. 1 the monotone measure $\hat{\mu}_{pan}$ is just the $\oplus$-optimal measure $\mu_\oplus$ to μ. Though, the optimal measure $\mu_\oplus$ and $\hat{\mu}_{pan}$ coincides, in order to compare with the case of the concave integral, in the following we still adopt the symbols used in the discussion of concave integrals (see [9]). We stress that the monotone measure $\hat{\mu}_{pan}^{(\oplus,\otimes)}$ generated by μ and the corresponding pan-integral does not depend on the pan-multiplication (Proposition 4).

Given a monotone measure $\mu \in \mathcal{M}$. A nonnegative set function $\hat{\mu}_{pan}^{(\oplus,\otimes)}$ generated by μ is defined as

$$\hat{\mu}_{pan}^{(\oplus,\otimes)}(A) = (p) \int^{(\oplus,\otimes)} \chi_A d\mu, \quad \forall A \in \mathcal{A}.$$

Then, it follows from Proposition 3 that $\hat{\mu}_{pan}^{(\oplus,\otimes)} \in \mathcal{M}$, and $\mu \leq \hat{\mu}_{pan}^{(\oplus,\otimes)}$, i.e., $\forall A \in \mathcal{A}$, $\mu(A) \leq \hat{\mu}_{pan}^{(\oplus,\otimes)}(A)$. Moreover, we will show that $\hat{\mu}_{pan}^{(\oplus,\otimes)}$ is totally balanced (for pan-integral), i.e., $\hat{\hat{\mu}}_{pan}^{(\oplus,\otimes)} = \hat{\mu}_{pan}^{(\oplus,\otimes)}$, that is,

$$\hat{\mu}_{pan}^{(\oplus,\otimes)}(A) = (p) \int^{(\oplus,\otimes)} \chi_A d\hat{\mu}_{pan}^{(\oplus,\otimes)}. \tag{3}$$

On the other hand, it follows from Proposition 4 that the monotone measure $\hat{\mu}_{pan}^{(\oplus,\otimes)}$ generated by μ is just the $\oplus$-optimal measure $\mu_\oplus$ to μ and $\oplus$, that is,

$$\hat{\mu}_{pan}^{(\oplus,\otimes)}(A) = \mu_\oplus(A) = (p) \int^{(\oplus,\otimes)} \chi_A d\mu.$$

Therefore, in order to prove $\hat{\hat{\mu}}_{pan}^{(\oplus,\otimes)} = \hat{\mu}_{pan}^{(\oplus,\otimes)}$, we only need to prove $\hat{\mu}_\oplus = \mu_\oplus$.

The following theorem is our main result.

Theorem 1. *Given a pan-space $(X, \mathcal{A}, \mu, \overline{R}_+, \oplus, \otimes)$. Then for all $f \in \mathcal{F}_+$,*

$$(p) \int^{(\oplus,\otimes)} f d\mu_\oplus = (p) \int^{(\oplus,\otimes)} f d\mu. \tag{4}$$

In particular, for each $A \in \mathcal{A}$, we have

$$(p) \int^{(\oplus,\otimes)} \chi_A d\mu_\oplus = (p) \int^{(\oplus,\otimes)} \chi_A d\mu, \tag{5}$$

i.e.,

$$(p) \int^{(\oplus,\otimes)} \chi_A d\mu_\oplus = \mu_\oplus(A), \tag{6}$$

that is $\hat{\mu}_\oplus = \mu_\oplus$.

In order to prove this theorem, we prepare two lemmas.

Lemma 1. *Given $(X, \mathcal{A}, \mu, \overline{R}_+, \oplus, \otimes)$. If $A, B \in \mathcal{A}$ and $A \cap B = \emptyset$, then*

$$(p) \int^{(\oplus,\otimes)} (\chi_A \oplus \chi_B) d\mu \geq (p) \int^{(\oplus,\otimes)} \chi_A d\mu \oplus (p) \int^{(\oplus,\otimes)} \chi_B d\mu. \tag{7}$$

Proof. Since the optimal measure $\mu_\oplus$ to μ is super-$\oplus$-additive (see Proposition 1), therefore

$$\mu_\oplus(A \cup B) \geq \mu(A) \oplus \mu(B). \tag{8}$$

On the other hand,

$$\mu_\oplus(A) = (p) \int^{(\oplus,\otimes)} \chi_A d\mu \quad \text{and} \quad \mu_\oplus(B) = (p) \int^{(\oplus,\otimes)} \chi_B d\mu, \tag{9}$$

noting that $\chi_A \oplus \chi_B = \chi_{A \cup B}$,

$$\mu_\oplus(A \cup B) = (p) \int^{(\oplus,\otimes)} (\chi_A \oplus \chi_B) d\mu, \tag{10}$$

combining (8), (9) and (10) then the result is obtained. $\qquad\square$

Lemma 2. *Given a pan-space $(X, \mathcal{A}, \mu, \overline{R}_+, \oplus, \otimes)$, for any pseudo-simple function*

$$s(x) = \bigoplus_{i=1}^{n} [a_i \otimes \chi_{A_i}(x)],$$

we have

$$(p) \int^{(\oplus,\otimes)} \left(\bigoplus_{i=1}^{n} [a_i \otimes \chi_{A_i}(x)] \right) d\mu \geq \bigoplus_{i=1}^{n} \left\{ (p) \int^{(\oplus,\otimes)} [a_i \otimes \chi_{A_i}(x)] d\mu \right\}$$

$$\geq \bigoplus_{i=1}^{n} \left\{ a_i \otimes \left((p) \int^{(\oplus,\otimes)} \chi_{A_i}(x) d\mu \right) \right\}.$$

Proof. We only discuss the case of $n = 2$ without loss of generality. Suppose $s(x) = (a \otimes \chi_A(x)) \oplus (b \otimes \chi_B(x))$, where $A \cap B = \emptyset$.

For any given $\epsilon > 0$, there are pseudo-simple functions $\bigoplus_{i=1}^{n} [\lambda_i \otimes \chi_{A_i}(x)]$ and $\bigoplus_{j=1}^{m} [\delta_j \otimes \chi_{B_j}(x)]$ such that

$$\bigoplus_{i=1}^{n} [\lambda_i \otimes \chi_{A_i}] \leq a \otimes \chi_A, \qquad \bigoplus_{j=1}^{m} [\delta_j \otimes \chi_{B_i}] \leq b \otimes \chi_B,$$

and

$$(p) \int^{(\oplus,\otimes)} \left(a \otimes \chi_A \right) d\mu < \bigoplus_{i=1}^{n} \left[\lambda_i \otimes \mu(A_i) \right] + \epsilon$$

and

$$(p) \int^{(\oplus,\otimes)} \left(b \otimes \chi_B \right) d\mu < \bigoplus_{j=1}^{m} \left[\delta_j \otimes \mu(B_j) \right] + \epsilon.$$

Where we can assume $\lambda_i > 0, i = 1, 2, \cdots, n$ and $\delta_j > 0, j = 1, 2, \cdots, m$ without loss of generality. Note that $\left(\bigoplus_{i=1}^{n} [\lambda_i \otimes \chi_{A_i}] \right) \oplus \left(\bigoplus_{j=1}^{m} [\delta_j \otimes \chi_{B_j}] \right)$ is a pseudo-simple function such that

$$\left(\bigoplus_{i=1}^{n} [\lambda_i \otimes \chi_{A_i}] \right) \oplus \left(\bigoplus_{j=1}^{m} [\delta_j \otimes \chi_{B_j}] \right) \leq (a \otimes \chi_A) \oplus (b \otimes \chi_B),$$

therefore

$$(p) \int^{(\oplus,\otimes)} \left(a \otimes \chi_A \right) d\mu \bigoplus (p) \int^{(\oplus,\otimes)} \left(b \otimes \chi_B \right) d\mu$$

$$< \bigoplus_{i=1}^{n} [\lambda_i \otimes \mu(A_i)] \bigoplus \bigoplus_{j=1}^{m} [\delta_j \otimes \mu(B_j)] + 2\epsilon$$

$$< (p) \int^{(\oplus,\otimes)} \left((a \otimes \chi_A) \oplus (b \otimes \chi_B) \right) d\mu + 2\epsilon.$$

Since ϵ is arbitrary and from Proposition 3 we have

$$(p) \int^{(\oplus,\otimes)} \left((a \otimes \chi_A) \oplus (b \otimes \chi_B) \right) d\mu$$

$$\geq (p) \int^{(\oplus,\otimes)} \left(a \otimes \chi_A \right) d\mu \bigoplus (p) \int^{(\oplus,\otimes)} \left(b \otimes \chi_B \right) d\mu$$

$$\geq \left(a \otimes (p) \int^{(\oplus,\otimes)} \chi_A d\mu \right) \bigoplus \left(b \otimes (p) \int^{(\oplus,\otimes)} \chi_B d\mu \right).$$

Similarly, for general pseudo-simple function $s(x)$ we can prove the conclusion of lemma.   □

Proof of Theorem 1. Obviously, we have the following inequality

$$(p) \int^{(\oplus,\otimes)} f d\mu_{\oplus} \geq (p) \int^{(\oplus,\otimes)} f d\mu. \tag{11}$$

Now we show that the opposite inequality also holds. We assume $(p) \int^{(\oplus,\otimes)} f d\mu_{\oplus} < +\infty$, in the case of $(p) \int^{(\oplus,\otimes)} f d\mu_{\oplus} = +\infty$, the proof is similar.

For any $\epsilon > 0$, there is pseudo-simple function $\bigoplus_{i=1}^{n} [a_i \otimes \chi_{A_i}(x)]$ such that $\bigoplus_{i=1}^{n} [a_i \otimes \chi_{A_i}] \leq f$ and

$$(p) \int^{(\oplus, \otimes)} f d\mu_\oplus < \bigoplus_{i=1}^{n} [a_i \otimes \mu_\oplus(A_i)] + \epsilon. \tag{12}$$

Combining Lemmas 1 and 2, and Proposition 3, we have

$$\bigoplus_{i=1}^{n} [a_i \otimes \mu_\oplus(A_i)] = \bigoplus_{i=1}^{n} \left\{ a_i \otimes \left((p) \int^{(\oplus, \otimes)} \chi_{A_i}(x) d\mu \right) \right\}$$

$$\leq \bigoplus_{i=1}^{n} \left\{ (p) \int^{(\oplus, \otimes)} [a_i \otimes \chi_{A_i}(x)] d\mu \right\}$$

$$\leq (p) \int^{(\oplus, \otimes)} \left(\bigoplus_{i=1}^{n} [a_i \otimes \chi_{A_i}(x)] \right) d\mu$$

$$\leq (p) \int^{(\oplus, \otimes)} f d\mu.$$

Therefore, it follows from formulation (12) that

$$(p) \int^{(\oplus, \otimes)} f d\mu_\oplus \leq (p) \int^{(\oplus, \otimes)} f d\mu + \epsilon. \tag{13}$$

Let $\epsilon \to 0$, then

$$(p) \int^{(\oplus, \otimes)} f d\mu_\oplus \leq (p) \int^{(\oplus, \otimes)} f d\mu \tag{14}$$

and hence

$$(p) \int^{(\oplus, \otimes)} f d\mu_\oplus = (p) \int^{(\oplus, \otimes)} f d\mu. \tag{15}$$

The proof of theorem is completed.   □

6   Conclusions

Given a pan-space $(X, \mathcal{A}, \mu, \overline{R}_+, \oplus, \otimes)$, we have discussed the relation between the pan-integral with respect to μ and the pan-integral with respect to $\mu_\oplus$. We have shown that these two pan-integrals (with respect to μ and $\mu_\oplus$, respectively) are coincident for all $f \in \mathcal{F}_+$ (Theorem 1). In particular, the monotone measure $\mu_\oplus$ generated by μ is totally balanced with respect to the pan-integrals. We stress that pan integrals for possibly two different measures μ and η coincide whenever $\mu_\oplus = \eta_\oplus$ [13]. As optimal measures coincide with super-$\oplus$-additive measures (Propositions 1 and 2), this means that we have an equivalence on measures, pan integrals coincide when measures are equivalent, and each equivalence class is represented by its unique super-$\oplus$-additive member (which is an optimal measure related to any of measures form the considered equivalence class), see [13].

When we discuss the properties related to the pan-integrals based on $(\oplus, \otimes)$, we only need to consider the case that the involved monotone measures are super-$\oplus$-additive.

Acknowledgements. This research was partially supported by the National Natural Science Foundation of China (Grant No. 11371332 and No. 11571106) and the NSF of Zhejiang Province (No. LY15A010013).

References

1. Benvenuti, P., Mesiar, R., Vivona, D.: Monotone set functions-based integrals. In: Pap, E. (ed.) Handbook of Measure Theory, vol. II, pp. 1329–1379. Elsevier, Amsterdam (2002)
2. Benvenuti, P., Mesiar, R.: Pseudo-arithmetical operations as a basis for the general measure and integration theory. Inf. Sci. **160**, 1–11 (2004)
3. Choquet, G.: Theory of capacities. Ann. Inst. Fourier **5**, 131–295 (1954)
4. Grabisch, M., Murofushi, T., Sugeno, M. (eds.): Fuzzy Measures and Integrals: Theory and Applications. Studies in Fuzziness and Soft Computing, vol. 40. Springer, Heidelberg (2000)
5. Ichihashi, H., Tanaka, M., Asai, K.: Fuzzy integrals based on pseudo-additions and multiplications. J. Math. Anal. Appl. **130**, 354–364 (1988)
6. Klement, E.P., Li, J., Mesiar, R., Pap, E.: Integrals based on monotone set functions. Fuzzy Sets Syst. **281**, 88–102 (2015)
7. Klement, E.P., Mesiar, R., Pap, E.: A universal integral as common frame for Choquet and Sugeno integral. IEEE Trans. Fuzzy Syst. **18**, 178–187 (2010)
8. Lehrer, E.: A new integral for capacities. Econ. Theor. **39**, 157–176 (2009)
9. Lehrer, E., Teper, R.: The concave integral over large spaces. Fuzzy Sets Syst. **159**, 2130–2144 (2008)
10. Li, J., Mesiar, R., Struk, P.: Pseudo-optimal measures. Inf. Sci. **180**, 4015–4021 (2010)
11. Mesiar, R.: Choquet-like integrals. J. Math. Anal. Appl. **194**, 477–488 (1995)
12. Mesiar, R., Li, J., Pap, E.: Pseudo-concave integrals. In: Li, S., Wang, X., Okazaki, Y., Kawabe, J., Murofushi, T., Guan, L. (eds.) NLMUA2011. Advances in Intelligent and Soft Computing, vol. 100, pp. 43–49. Springer, Heidelberg (2011)
13. Mesiar, R., Li, J., Pap, E.: Discrete pseudo-integrals. Int. J. Approx. Reason. **54**, 357–364 (2013)
14. Mesiar, R., Rybárik, J.: Pan-operaions structure. Fuzzy Sets Syst. **74**, 365–369 (1995)
15. Mesiar, R., Stupnaňová, A.: Decomposition integrals. Int. J. Approx. Reason. **54**, 1252–1259 (2013)
16. Ouyang, Y., Li, J.: An equivalent definition of pan-integral. In: Torra, V., Narukawa, Y., Navarro-Arribas, G., Yañez, C. (eds.) MDAI 2016. LNCS, vol. 9880, pp. 107–113. Springer, Cham (2016). doi:10.1007/978-3-319-45656-0_9
17. Pap, E.: Null-Additive Set Functions. Kluwer, Dordrecht (1995)
18. Shilkret, N.: Maxitive measure and integration. Indag. Math. **33**, 109–116 (1971)
19. Sugeno, M.: Theory of fuzzy integrals and its applications. Ph.D. Dissertation, Takyo Institute of Technology (1974)
20. Sugeno, M., Murofushi, T.: Pseudo-additive measures and integrals. J. Math. Anal. Appl. **122**, 197–222 (1987)

21. Wang, Z., Klir, G.J.: Generalized Measure Theory. Springer, New York (2009)
22. Wang, Z., Wang, W., Klir, G.J.: Pan-integrals with respect to imprecise probabilities. Int. J. Gen Syst **25**, 229–243 (1996)
23. Yang, Q.: The pan-integral on fuzzy measure space. Fuzzy Math. **3**, 107–114 (1985). (in Chinese)
24. Zhang, Q., Mesiar, R., Li, J., Struk, P.: Generalized Lebesgue integral. Int. J. Approx. Reason. **52**, 427–443 (2011)

Orness and Cardinality Indices for Averaging Inclusion-Exclusion Integrals

Aoi Honda[1], Simon James[2]([⊠]), and Sutharshan Rajasegarar[2]

[1] Kyushu Institute of Technology, Iizuka, Japan
aoi@ces.kyutech.ac.jp
[2] Deakin University, Burwood, Australia
{sjames,srajas}@deakin.edu.au

Abstract. The inclusion-exclusion integral is a generalization of the discrete Choquet integral, defined with respect to a fuzzy measure and an interaction operator that replaces the minimum function in the Choquet integral's Möbius representation. While in general this means that the resulting operator can be non-monotone, we have previously proposed using averaging aggregation functions for the interaction component, which under certain requirements can produce non-linear, but still averaging, operators. Here we consider how the orness of the overall function changes depending on the chosen component functions and hence propose a simplified calculation for approximating the orness of an averaging inclusion-exclusion integral.

1 Introduction

Understanding interaction amongst variables is not easy. An important problem in data analysis and supervised learning is how to determine the importance of different predictor variables, however in addition to overall importance, it may be that variables have different effects in different parts of the domain or that they interact in some way.

The Choquet integral [1,2] has been studied in the context of decision making for over 20 years and is useful in terms of its ability to model interaction effects while also being relatively easy to interpret. On this latter virtue, the ability to make use of a function like the Choquet integral in data analysis is predominantly thanks to the dedicated works that have developed alternative representations, measures of importance, interaction indices, and so forth (e.g. [3,4]). In this work, we focus on the orness [5] and cardinality indices [6] and how they can be extended to the *inclusion-exclusion integral*, a generalization of the Choquet integral that was proposed in [7], with further studies in e.g. [8].

As well as using a fuzzy measure in its definition, the inclusion-exclusion integral (IE-integral) employs an interaction operator $I(\mathbf{x}|A)$ defined for $2, \ldots, n$ variables. It can be interpreted as breaking down the inputs into positive and negative interacting contributions in an analogous way to the Möbius representation of a fuzzy measure. In some cases this will result in non-monotone operators. In [9] the idea of using averaging functions for $I(\mathbf{x}|A)$ was proposed

© Springer International Publishing AG 2017
V. Torra et al. (Eds.): MDAI 2017, LNAI 10571, pp. 51–62, 2017.
DOI: 10.1007/978-3-319-67422-3_6

and conditions on ensuring that the resulting IE-integral is itself averaging were investigated.

One interesting observation was that as the $I(\mathbf{x}|A)$ components graduate between the minimum and maximum, the resulting IE-integral graduates between the Choquet integral with respect to its fuzzy measure μ and the dual Choquet integral with respect to μ^*. The Shapley indices, which indicate the average importance assigned to each variable by μ are unchanged under duality, however the orness (the degree to which the function behaves like the maximum) of μ will be complementary to that of μ^*, i.e. $\mathsf{orness}(\mu) = 1 - \mathsf{orness}(\mu^*)$. This means that the orness of the IE-integral will also change with $I(\mathbf{x}|A)$, as will the cardinality indices [6,10], which give the average importance of inputs based on their relative ranking.

In this contribution we will establish that, while the orness of averaging inclusion-exclusion integrals may be difficult to calculate in general, a suitable approximation can be obtained based on the orness associated with μ and the orness of the component function $I(\mathbf{x}|A)$ when $|A| = n$.

The paper will be set out as follows. In the Preliminaries section we will give the necessary background on aggregation functions and the Choquet integral. We then proceed to give an overview of the inclusion-exclusion integral with respect to averaging functions in Sect. 3 before defining our orness approximation. In Sect. 4, we present some results regarding the use of OWA functions for $I(\mathbf{x}|A)$ before providing some examples and calculations in Sect. 5. We briefly conclude in Sect. 6.

2   Preliminaries

This contribution concerns orness and cardinality indices, which have been used for years to characterize the behavior of averaging aggregation functions. We will hence first recall a number of definitions relating to aggregation functions and, in particular, the Choquet integral. The inclusion-exclusion integral can be understood as a generalization of the Choquet integral, and therefore the indices used to define orness with respect to fuzzy measures will be particularly relevant for us.

2.1   Aggregation Functions

When it comes to modeling decisions, aggregation functions (e.g. see [11–13]) will inevitably play some role in either summarizing data or making overall evaluations.

We will contain ourselves to averaging aggregation functions defined over $[0,1]^n$, although in general, many results will apply to any interval.

Definition 1 (Aggregation function). *For $n \in \mathbb{N}\backslash\{1\}$ inputs given over the real interval $[0,1]^n$, an aggregation function $\mathsf{F} : [0,1]^n \to [0,1]$ is a function non-decreasing in each argument and satisfying $\mathsf{F}(0,\ldots,0) = 0$ and $\mathsf{F}(1,\ldots,1) = 1$.*

The focus of this work is orness, which relates to *averaging* aggregation functions. An aggregation function F is averaging when for all $\mathbf{x} \in [0,1]^n$,

$$\min(\mathbf{x}) \leq F(\mathbf{x}) \leq \max(\mathbf{x}).$$

The monotonicity, or non-decreasingness, of aggregation functions means that averaging behavior ensures idempotency (and idempotency ensures averaging behavior). Typical averaging functions include the weighted arithmetic means (WAMs) and ordered weighted averaging (OWA) operators. For a weighting vector $\mathbf{w}$, such that $\sum_{i=1}^n w_i = 1$ and $w_i \geq 0$ for all i, the weighted arithmetic mean (WAM) is given by $\mathsf{WAM}(\mathbf{x}) = \sum_{i=1}^n w_i x_i.$, while the OWA [14] is $\mathsf{OWA}(\mathbf{x}) = \sum_{i=1}^n w_i x_{(i)}.$, where $x_{(i)}$ denotes the $i-th$ largest of the inputs.

The weighted arithmetic mean takes the weighted sum of the inputs, usually allocating w_i based on the importance of the i-th criterion or attribute of the dataset, while the OWA operator assigns w_i to the i-th largest input.

The OWA operator and WAM will coincide when $w_i = 1/n$ for all i (i.e. an unweighted arithmetic mean). The OWA also includes special cases such as the maximum when $w_1 = 1$, the minimum if $w_n = 1$, and the median if the middle weight is equal to 1 (or spread across the middle two weights if n is even).

The quasi-arithmetic means generalize the weighted arithmetic means to model non-linear relationships using a generating function g. With respect to these invertible and monotone functions $g : [0,1] \to [-\infty, \infty]$ the weighted quasi-arithmetic mean is given by

$$M_g(\mathbf{x}) = g^{-1}\left(\sum_{i=1}^n w_i g(x_i)\right).$$

The weighted power means ($M_p(\mathbf{x})$) are obtained if $g(t) = t^p$, including the quadratic mean QM when $p = 2$. The geometric mean $GM(\mathbf{x}) = \prod_{i=1}^n x_i^{w_i}$ can be obtained using $g(t) = \ln t$ and is also a limiting case for the power means as $t \to 0$. Of course, $g(t) = t$ returns the WAM.

We will also refer to the *dual* of an aggregation function. The dual F^d of any aggregation function F can be obtained using the standard negation $N(t) = 1-t$ and the construction,

$$F^d(\mathbf{x}) = 1 - F(1 - x_1, 1 - x_2, \ldots, 1 - x_n).$$

The dual of the minimum function is the maximum and vice-versa. Weighted arithmetic means are self-dual and for a given quasi-arithmetic mean with generator $g(t)$, its dual function with respect to the standard negation $N(t) = 1-t$ is generated by $g(1-t)$.

In general, if a function F tends more toward lower inputs, its dual F^d will tend more towards higher inputs. For an OWA function with respect to $\mathbf{w}$, the dual function will be an OWA with the same weights in reverse order, e.g. if $\mathbf{w} = \langle 0.3, 0.4, 0.2, 0.1 \rangle$ then the weighting vector of the dual will be $\mathbf{w}^d = \langle 0.1, 0.2, 0.4, 0.3 \rangle$.

2.2   The Choquet Integral

The Choquet integral is a particularly expressive function that generalizes both the OWA and WAM in a different manner to the quasi-arithmetic means. Rather than a weighting vector $\mathbf{w}$, the Choquet integral uses a fuzzy measure μ, which can be interpreted as a set function that allocates weight to every subset of inputs.

Definition 2 (Fuzzy Measure). *For a given finite set* $\{1 : n\} = \{1, 2, \ldots, n\}$, *a fuzzy measure is a set function* μ *defined for all* $S \subseteq \{1 : n\}$ *such that* $\mu(\emptyset) = 0, \mu(\{1 : n\}) = 1$ *and* $S \subseteq T$ *implies* $\mu(S) \leq \mu(T)$.

The discrete Choquet integral for a finite set of inputs can then be expressed as

$$C_\mu(\mathbf{x}) = \sum_{i=1}^{n} x_{(i)} \Big(\mu(\{(i) : (n)\}) - \mu(\{(i+1) : (n)\}) \Big),$$

where the notation $x_{(i)}$ means the inputs are re-arranged in non-decreasing order and $\{(i) : (n)\}$ refers to the set of (ordered) inputs from the (i)-th smallest to the largest and $\{(i + 1) : (n)\}$ is the set of the $(i + 1)$-th smallest to the largest.

The Choquet integral can be given in terms of the Möbius transform of the fuzzy measure. This can sometimes simplify computations and the learning of weight parameters from data. The Möbius representation m is calculated from μ as follows.

$$m_A = \sum_{B \subseteq A} (-1)^{|A \setminus B|} v(B).$$

The Choquet integral is then given by $C_\mu(\mathbf{x}) = C_m = \sum_{A \subseteq N}^{n} m_A \min(\mathbf{x}|A)$, where $\min(\mathbf{x}|A)$ indicates the minimum x_i such that $i \in A$.

Since the number of subsets grows exponentially with n, the Shapley value [15] has been used to measure the average importance of each variable [2,3]. It is denoted by the vector

$$\boldsymbol{\phi} = \langle \phi_1, \phi_2, \ldots, \phi_n \rangle,$$

and can be calculated directly from the fuzzy measure μ where

$$\phi_i = \sum_{A \subseteq N \setminus \{i\}} \frac{(n - |A| - 1)!|A|!}{n!} (\mu(A \cup \{i\}) - \mu(A)).$$

It can be interpreted similarly[1] to the weighting vector $\mathbf{w}$.

The dual fuzzy measure μ^*, which defines the dual Choquet integral, can be obtained using $\mu^*(A) = 1 - \mu(A')$ where A' is the complement of A. The Shapley index remains the same under duality.

---

[1] It has been noted in [16] that the least squares linear approximation of a given function f (which could be used to infer the importance of each variable) actually corresponds with the Banzhaf index, a calculation similar to the Shapley index.

Another index proposed in [6] and more recently applied in [17] and [10] is the cardinality index, which allows the average order weights (or rank weights) of a fuzzy measure to be characterized. For a given fuzzy measure μ, the cardinality index ϑ_i for each i in $\{1, 2, \ldots, n\}$ is given by

$$
\vartheta_i = \left(\frac{1}{\binom{n}{i}} \sum_{|A|=i} \mu(A) \right) - \left(\frac{1}{\binom{n}{i-1}} \sum_{|A|=i-1} \mu(A) \right).
$$

While the Shapley index should roughly correspond with the weights of a WAM, the cardinality indices should be viewed similarly to the weights of an OWA.

2.3   Orness

The measure of orness is intended to characterize averaging aggregation functions in terms of how close they are, on average, to the maximum operator, i.e., their tendency toward the higher inputs or the degree to which they behave *disjunctively*. The complementary calculation 'andness' is the orness subtracted from 1 and gives the degree to which the function behaves *conjunctively*.

The following definition was proposed by Dujmovic in 1973 [5,18]. We will mainly be concerned with calculations over the unit hypercube $[0,1]^n$, however it will be useful to provide the following calculation formulas in terms of the general interval $[a, b]$.

For an averaging aggregation function F, the measure of orness is given by

$$
\mathsf{orness}(\mathsf{F}) = \frac{\int_{[a,b]^n} \mathsf{F}(\mathbf{x}) \, \mathrm{d}\mathbf{x} - \int_{[a,b]^n} \min(\mathbf{x}) \, \mathrm{d}\mathbf{x}}{\int_{[a,b]^n} \max(\mathbf{x}) \, \mathrm{d}\mathbf{x} - \int_{[a,b]^n} \min(\mathbf{x}) \, \mathrm{d}\mathbf{x}}.
$$

When F is the maximum, we have $\mathsf{orness}(\mathsf{F}) = 1$ while F being the minimum results in an orness of 0.

The integral of F is often done numerically, however this too can be computationally expensive for large n. Some functions have closed form solutions, and so we will find the following results useful [11].

$$
\int_{[a,b]^n} \max(\mathbf{x}) \, \mathrm{d}\mathbf{x} = (b - a)^n \frac{a + bn}{n + 1}, \quad \int_{[a,b]^n} \min(\mathbf{x}) \, \mathrm{d}\mathbf{x} = (b - a)^n \frac{an + b}{n + 1} \quad (1)
$$

The orness of the geometric mean is given with respect to n by

$$
\mathsf{orness}(\mathsf{G}) = -\frac{1}{n - 1} + \frac{n + 1}{n - 1} \left(\frac{n}{n + 1} \right)^n, \tag{2}
$$

however for other well known averaging functions, only some special cases have such calculation formulas.

Approximation formulas have been proposed for the quasi-arithmetic means, in particular we recall those of Dujmovic (orness$_D$) and Liu (orness$_L$) [19].

$$\text{orness}_D(M_g) = \frac{g(b) - \int_a^b g(t)\,dt}{g(b) - g(a)}, \quad \text{orness}_L(M_g) = \frac{g^{-1}\left(\frac{\int_a^b g(t)\,dt}{b-a}\right) - a}{b - a}. \tag{3}$$

The true orness will always be bound between the two calculations. The orness$_L$ will approach the true value for large n, while orness$_D$ represents a kind of limiting case when $n = 1$. Both values may differ markedly from the real value, e.g. when $n = 2$, the orness of the quadratic mean is approximately 0.6232 while Liu's formula yields 0.5774 and Dujmovic's gives 0.6667 (correct to 4 decimal places).

For the OWA and Choquet integral, the exact orness values can be calculated directly from the weights and fuzzy measure respectively. For the OWA function [14],

$$\text{orness}(\mathbf{w}) = \sum_{i=1}^{n} w_i \frac{n - i}{n - 1}, \tag{4}$$

while for the Choquet integral the orness is given by [20],

$$\text{orness}(\mu) = \frac{1}{n - 1} \sum_{A \subset \mathbb{N}} \frac{(n - |A|)! |A|!}{n!} \mu(A). \tag{5}$$

Note that $A \subset \mathbb{N}$ should be strict, so that $\mathbb{N}$ is not included in the sum. The orness for the Choquet integral could also be calculated from the cardinality indices using the OWA orness calculation.

3   Defining Orness for the Inclusion-Exclusion Integral

The indices for interpreting the Choquet integral's behavior facilitate its use in regression and data analysis, since the Shapley indices can be used to infer the importance of variables and the orness and cardinality indices allow us to interpret the overall behavior in terms of tendency toward higher or lower inputs. The inclusion-exclusion integral (IE-integral) generalizes the discrete Choquet integral by means of a function $I(\mathbf{x}|A)$, referred to as the interaction operator.

For this work, we consider $I(\mathbf{x}|A)$ to be a family of averaging functions[2] with $\mathbf{x}|A$ denoting the input vector $\mathbf{x}$ restricted to the inputs x_i such that $i \in A$, defined for each $|A| \in \{1 : n\}$ and satisfying projection, i.e. $I(\mathbf{x}|i) = x_i$. From this we obtain the following definition for the inclusion-exclusion integral [7].

---

[2] We note however that in [7,8], $I(\mathbf{x}|A)$ is proposed to be arity-decreasing, i.e. $I(\mathbf{x}|A) < I(\mathbf{x}|B)$ if $B \subset A$. As noted in [9], instead using averaging functions (which are not necessarily arity-decreasing) will allow the IE-integral to be averaging.

Definition 3 (Inclusion-Exclusion integral). *For a fuzzy measure μ and an interaction operator I, the inclusion-exclusion (IE) integral is given by,*

$$IE_{I,\mu}(\mathbf{x}) = \sum_{A \subseteq N} \mu(A) M^I(A)$$

where $M^I(A) = \sum_{B \supseteq A} (-1)^{|B \setminus A|} I(\mathbf{x}|B)$.

The IE-integral can also be expressed with respect to the Möbius transformation of the defining fuzzy measure,

$$IE_{I,m}(\mathbf{x}) = \sum_{A \subseteq N} m_A I(\mathbf{x}|A).$$

It has been established in [9] that any μ with Möbius values greater than zero along with $I(\mathbf{x}|A)$ being averaging will result in averaging IE-integrals, as will any $I(\mathbf{x}|A)$ whose partial derivatives decrease sufficiently with increases to arity (regardless of μ).

Different choices of $I(\mathbf{x}|A)$ have the effect of defining functions that graduate between a discrete Choquet integral with respect to μ and a Choquet integral with respect to the dual fuzzy measure μ^* (e.g. See Fig. 1).

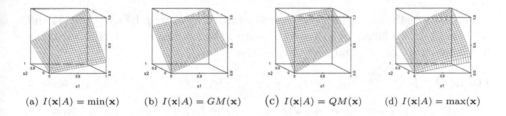

(a) $I(\mathbf{x}|A) = \min(\mathbf{x})$ (b) $I(\mathbf{x}|A) = GM(\mathbf{x})$ (c) $I(\mathbf{x}|A) = QM(\mathbf{x})$ (d) $I(\mathbf{x}|A) = \max(\mathbf{x})$

Fig. 1. Inclusion-exclusion integrals with respect to different interaction operators I.

Even for averaging functions defined in a straightforward way for all n (e.g. the quasi-arithmetic means), the orness will vary depending on the arity. This makes calculating the orness of the IE-integral problematic, and so we propose an approximation of the overall orness using the following calculation. We denote the orness of the Choquet integral with respect to μ by $\Omega(\mu)$ and an approximation or calculation of orness for $I(\mathbf{x}|\{1 : n\})$ by $\Omega(I)$. Then,

$$\text{orness}(IE) \approx (1 - \Omega(I)) \cdot \Omega(\mu) + \Omega(I) \cdot (1 - \Omega(\mu)). \tag{6}$$

Similarly we can approximate the cardinality indices:

$$\vartheta_i = (1 - \Omega(I)) \cdot \vartheta_i + \Omega(I) \cdot \vartheta_{n-i+1}. \tag{7}$$

In the following section we will highlight some special cases where $I(\mathbf{x}|A)$ is chosen to be a set of OWA operators before providing some more general examples and their orness calculations.

4   IE-integrals with Respect to OWA Operators

When a set of OWA operators are used for the interaction operator $I(\mathbf{x}|A)$, i.e., OWAs defined for $n = 2, \ldots, n$, the IE-integral will be a weighted sum (with respect to not necessarily positive weights). This is because once the ordering of the inputs is fixed, the weights assigned to each variable from the OWA operators of differing dimension will be set and the IE-integral can be calculated as a linear multiple of these and the Möbius weights. In some cases, this will simply coincide with a Choquet integral with respect to an alternative fuzzy measure. We consider OWAs here in order to help show the relationship between orness and the approximation equation.

4.1   Case of $n = 2$ Where I Is an OWA

We can show that Eq. (6) will hold exactly when I is an OWA and $n = 2$.

Proposition 1. *For a 2-variate inclusion-exclusion integral with respect to a fuzzy measure μ and $I(\mathbf{x}|A)$, let $\Omega(\mu)$ denote the orness associated with μ and let $I(\mathbf{x}|A)$ be given by an OWA operator with weights w_1, w_2. The resulting function will be equivalent to a Choquet integral with respect to a fuzzy measure v and orness given by $w_2 \cdot \Omega(\mu) + w_1 \cdot (1 - \Omega(\mu))$.*

Proof. Denote the singleton measures of the fuzzy measure μ by $\mu_1 = \mu(\{1\})$ and $\mu_2 = \mu(\{2\})$, the larger of x_1, x_2 by $x_{(1)}$ and the smaller by $x_{(2)}$. The resulting inclusion-exclusion integral can be expressed,

$$IE_{\mu,\mathbf{w}}(\mathbf{x}) = \mu_1(x_1 - w_1 x_{(1)} - w_2 x_{(2)}) + \mu_2(x_2 - w_1 x_{(1)} - w_2 x_{(2)}) + w_1 x_{(1)} + w_2 x_{(2)}.$$

If $x_1 > x_2$, then the coefficient of x_1 (the larger value) will be equal to $\mu_1(1 - w_1) + (1 - \mu_2)w_1$, which will correspond with the singleton of the resulting Choquet integral's fuzzy measure v, i.e. $v(\{1\})$. On the other hand, if $x_2 > x_1$, the coefficient of x_2 will be $\mu_2(1 - w_1) + (1 - \mu_1)w_1$, which denotes $v(\{2\})$.

Taking the average of these gives the orness calculation of v, $(1 - w_1)\frac{\mu_1 + \mu_2}{2} + w_1\left(\frac{2 - \mu_1 - \mu_2}{2}\right)$, which, since the orness of a 2-variate OWA is given by w_1 and the orness of a 2-variate Choquet integral is $\frac{\mu_1 + \mu_2}{2}$, corresponds with Eq. (6) and completes the proof.

In [9] it was established that 2-variate IE-integrals will be averaging if the function $I(\mathbf{x}|A)$ is 1-Lipschitz. Since this is always true for 2-variate OWA operators, using an OWA will always return a function that is equivalent to a Choquet integral. In the 3-variate case it is slightly more complicated, since use of an OWA for $I(\mathbf{x}|A)$ in 2- and 3-dimensions does not guarantee averaging behavior.

4.2   Case of $n = 3$ Where I_2 and I_3 Are OWA Operators

For the case of 3 variables, we need to define the interaction operator for 2- and 3-dimensions. While the quasi-arithmetic means have a straightforward extension, this is not the case with the OWA. As established in [9], we can ensure averaging behavior of the resulting IE-integral if the following inequalities hold.

$$1 + I'_i(x_1, x_2, x_3) - \sum_{j \in \mathbb{N} \backslash i} I'_i(x_i, x_j) > 0, \quad -I'_i(x_1, x_2, x_3) + \sum_{j \in \mathbb{N} \backslash i} I'_i(x_i, x_j) > 0,$$

for each i, where I'_i denotes the partial derivative with respect to i, i.e. the OWA weight associated with x_i.

Unlike the case of 2 variables, using OWA functions for IE-integrals of 3 variables will not always return a Choquet integral or averaging function (even those that are related by a consistent rule, e.g. quantifiers [21]). However we can show that the orness calculation will hold whenever the two OWA operators used have the same orness for 2 and 3 dimensions.

Proposition 2. *For a 3-variate inclusion-exclusion integral with respect to a fuzzy measure μ, denote the orness associated with μ by $\Omega(\mu)$ and let $I(\mathbf{x}|A)$ be defined by two OWA operators (for 2- and 3-dimensions) with the same orness $\Omega(OWA)$. The resulting function will have an orness given by*

$$(1 - \Omega(OWA)) \cdot \Omega(\mu) + \Omega(OWA) \cdot (1 - \Omega(\mu)).$$

Proof. Let $I(\mathbf{x}|A) = OWA_2(\mathbf{x}) = w_1^2 x_{(1)} + w_2^2 x_{(2)}$ when $|A| = 2$ and $I(\mathbf{x}|A) = OWA_3(\mathbf{x}) = w_1^3 x_{(1)} + w_2^3 x_{(2)} + w_3^3 x_{(3)}$ when $|A| = 3$.

As previously, We can consider the overall weight that will be applied to a particular input depending on its position, then take the average across the first position, and then the average across the second position so that we can calculate the corresponding orness.

Recall that in the calculation of the IE-integral, for each set A we have the term,

$$\mu(A) \sum_{B \supseteq A} (-1)^{|B \backslash A|} I(\mathbf{x}|B).$$

If x_1 is the largest input, its coefficient will be

$$\mu_1(1 - w_1^2 - w_1^2 + w_1^3) + \mu_2(-w_1^2 + w_1^3) + \mu_3(-w_1^2 + w_1^3)$$
$$+ \mu_{12}(w_1^2 - w_1^3) + \mu_{13}(w_1^2 - w_1^3) + \mu_{23}(-w_1^3) + \mu_{123} w_1^3.$$

If it is the second largest, and $(x_2 > x_1 > x_3)$, its coefficient becomes

$$\mu_1(1 - w_2^2 - w_1^2 + w_2^3) + \mu_2(-w_2^2 + w_2^3) + \mu_3(-w_1^2 + w_2^3)$$
$$+ \mu_{12}(w_2^2 - w_2^3) + \mu_{13}(w_1^2 - w_2^3) + \mu_{23}(-w_2^3) + \mu_{123} w_2^3,$$

while for $(x_3 > x_1 > x_2)$,

$$\mu_1(1 - w_1^2 - w_2^2 + w_2^3) + \mu_2(-w_1^2 + w_2^3) + \mu_3(-w_2^2 + w_2^3)$$
$$+ \mu_{12}(w_1^2 - w_2^3) + \mu_{13}(w_2^2 - w_2^3) + \mu_{23}(-w_2^3) + \mu_{123} w_2^3.$$

With the symmetric cases for x_2, x_3, the average weight allocated to the largest input is hence

$$\frac{\mu_1 + \mu_2 + \mu_3}{3}(1 - 4w_1^2 + 3w_1^3) + \frac{\mu_{12} + \mu_{13} + \mu_{23}}{3}(2w_1^2 - 3w_1^3) + w_1^3.$$

The average weight allocated to the second largest input is then

$$\frac{\mu_1 + \mu_2 + \mu_3}{6}(2 - 4w_1^2 - 4w_2^2 + 6w_2^3) + \frac{\mu_{12} + \mu_{13} + \mu_{23}}{6}(2w_1^2 + 2w_2^2 - 6w_2^3) + w_2^3.$$

Now, noting that $w_1^2 + w_2^2 = 1$, we use $\vartheta_1 = \frac{\mu_1 + \mu_2 + \mu_3}{3}$ to denote the average singleton weight of the fuzzy measure μ. We also have that the first two cardinality indices will be equal to the average weight of pairs, i.e. $\vartheta_1 + \vartheta_2 = \frac{\mu_{12} + \mu_{13} + \mu_{23}}{3}$. Using the formula for orness calculation of three variables, adding half of the second weight's average value to the first gives

$$\vartheta_1 \frac{2 - 16w_1^2 + 12w_1^3 + 6w_2^3}{4} + (\vartheta_1 + \vartheta_2)\frac{2 + 8w_1^2 - 12w_1^3 - 6w_2^3}{4} + w_1^3 + (1/2)w_2^3.$$

Since the OWA operators have the same orness, it will hold that $w_1^2 = w_1^3 + (1/2)w_2^3$. We then have

$$\vartheta_1 \frac{1 - 2w_1^2}{2} + (\vartheta_1 + \vartheta_2)\frac{1 - 2w_1^2}{2} + w_1^2 = w_1^2 + \vartheta_1 + \frac{\vartheta_2}{2} - 2\vartheta_1 w_1^2 - \vartheta_2 w_1^2$$

$$= (1 - w_1^2)(\vartheta_1 + \frac{\vartheta_2}{2}) + w_1^2(1 - \vartheta_1 - \frac{\vartheta_2}{2}).$$

Which corresponds with Eq. (6) since the orness of the OWA is w_1^2 and the orness of μ is $\vartheta_1 + \frac{\vartheta_2}{2}$.

5   Orness Calculations for Various IE-integrals

We now consider some examples of inclusion-exclusion integrals and calculate their orness numerically and using Eq. (6) to give an idea of the precision of the different approximations. For the numerical calculation we evaluated each function at approximately 10 million equally spaced points in the unit hypercube. In general this will be close to the true orness, although for 5-dimensions it may deviate significantly if the function has steep derivatives (e.g. for the geometric mean the true orness when $n = 5$ is 0.3528 to 4 decimal places whereas our numeric calculation gives 0.3562). In Table 1 we have summarized the calculations where $I(\mathbf{x}|A)$ is chosen to be either the geometric mean, a power mean with $p = 2$, or a power mean with $p = 5$. In order to define some example fuzzy measures, we learned the values from the BikeShare dataset [22] with 2, 3, 4, and 5 predictor variables (chosen from season, workday, weather, temperature, humidity) with the values optimized[3] with respect to a Choquet integral to predict the number of casual users.

---

[3] Using linear programming techniques as found, e.g. in [11]. Full details of the transformations and code used to learn fuzzy measures can be found at http://aggregationfunctions.wordpress.com.

Table 1. Approximated orness values for IE-integrals

n	orness(μ)	GM		PM ($p = 2$)			PM ($p = 5$)		
		num	app	num	Liu	Duj	num	Liu	Duj
2	0.220	0.407	0.406	0.569	0.543	0.594	0.663	0.612	0.687
3	0.186	0.402	0.402	0.569	0.549	0.605	0.669	0.625	0.710
4	0.176	0.403	0.402	0.566	0.550	0.608	0.665	0.629	0.716
5	0.208	0.416	0.414	0.556	0.545	0.597	0.644	0.616	0.695

The top spanning header row reads: $I(\mathbf{x}|A)$

As might have been expected, the approximated values using the Liu and Dujmovic calculations for power means deviated from the numerical calculations, while those of the geometric mean remained close. As an example of the change in cardinality indices, for $n = 4$, the cardinality indices of the original fuzzy measure are $\vartheta = \langle 0.01, 0.09, 0.16, 0.21, 0.53 \rangle$ whereas with PM and $p = 5$ this becomes $\vartheta = \langle 0.40, 0.18, 0.16, 0.12, 0.14 \rangle$ based on the numerical calculation of the power mean's orness (not using the approximated formulas). The use of $I(\mathbf{x}|A)$ can hence also be seen as a method of adjusting the cardinality indices for a given Choquet integral without affecting the Shapley index.

6 Conclusion

We have introduced a simple orness calculation that can be used to quickly evaluate or estimate the orness of an averaging IE-integral if the orness of the components, μ and $I(\mathbf{x}|A)$ are known. This helps enable interpretation of the inclusion-exclusion integral when it is used as an averaging generalization of the Choquet integral, however we note that in its original definition, $I(\mathbf{x}|A)$ was proposed to be arity-increasing and therefore the use of averaging functions produces a function with different properties than the original integral. While here we have used the orness calculation that is often referred to as the 'global orness', the results could also be compared with other indices such as the mean 'local orness' index (as given in [12]).

References

1. Choquet, G.: Theory of capacities. Ann. Inst. Fourier **5**, 131–295 (1953)
2. Grabisch, M.: The applications of fuzzy integrals in multicriteria decision making. Eur. J. Oper. Res. **89**, 445–456 (1996)
3. Grabisch, M.: k-order additive discrete fuzzy measures and their representation. Fuzzy Sets Syst. **92**, 167–189 (1997)
4. Grabisch, M., Kojadinovic, I., Meyer, P.: A review of methods for capacity identification in choquet integral based multi-attribute utility theory applications of the Kappalab R package. Eur. J. Oper. Res. **186**, 766–785 (2008)

5. Dujmovic, J.: Two integrals related to means, pp. 231–232. Univ. Beograd. Publ. Elektrotechn. Fak. (1973)
6. Yager, R.R.: On the cardinality index and attitudinal character of fuzzy measures. Int. J. Gen Syst **31**(3), 303–329 (2002)
7. Honda, A., Okamoto, J.: Inclusion-exclusion integral and its application to subjective video quality estimation. In: Hüllermeier, E., Kruse, R., Hoffmann, F. (eds.) IPMU 2010. CCIS, vol. 80, pp. 480–489. Springer, Heidelberg (2010). doi:10.1007/978-3-642-14055-6_50
8. Honda, A., Okazaki, Y.: Inclusion-exclusion integral and t-norm based data analysis model construction. In: Carvalho, J.P., Lesot, M.-J., Kaymak, U., Vieira, S., Bouchon-Meunier, B., Yager, R.R. (eds.) IPMU 2016. CCIS, vol. 610, pp. 65–77. Springer, Cham (2016). doi:10.1007/978-3-319-40596-4_7
9. Honda, A., James, S.: Averaging aggregation functions based on inclusion-exclusion integrals. In: Proceedings of the Joint World Congress of International Fuzzy Systems Association and International Conference on Soft Computing and Intelligent Systems, Otsu, Japan, pp. 1–6 (2017)
10. Labreuche, C.: On capacities characterized by two weight vectors. In: Carvalho, J.P., Lesot, M.-J., Kaymak, U., Vieira, S., Bouchon-Meunier, B., Yager, R.R. (eds.) IPMU 2016. CCIS, vol. 610, pp. 23–34. Springer, Cham (2016). doi:10.1007/978-3-319-40596-4_3
11. Beliakov, G., Bustince Sola, H., Calvo Sánchez, T.: A Practical Guide to Averaging Functions. SFSC, vol. 329. Springer, Cham (2016). doi:10.1007/978-3-319-24753-3
12. Grabisch, M., Marichal, J.-L., Mesiar, R., Pap, E.: Aggregation Functions. Cambridge University Press, Cambridge (2009)
13. Torra, V., Narukawa, Y.: Modeling Decisions. Information Fusion and Aggregation Operators. Springer, Heidelberg (2007)
14. Yager, R.: On ordered weighted averaging aggregation operators in multicriteria decision making. IEEE Trans. Syst. Man Cybern. **18**, 183–190 (1988)
15. Shapley, L.S.: A value for n-person games. In: Kuhn, H., Tucker, A. (eds.) Contributions to the Theory of Games, Annals of Mathematics Studies, vol. II, No. 28, pp. 307–317. Princeton University Press (1953)
16. Marichal, J.-L., Mathonet, P.: Measuring the interactions among variables of functions over the unit hypercube. J. Math. Anal. Appl. **380**, 105–116 (2011)
17. Marichal, J.L., Mathonet, P.: Symmetric approximations of pseudo-boolean functions with applications to influence indexes. Appl. Math. Lett. **25**, 1121–1126 (2012)
18. Dujmovic, J.: Weighted conjunctive and disjunctive means and their application in system evaluation, pp. 147–158. Univ. Beograd. Publ. Elektrotechn. Fak. (1974)
19. Liu, X.: An orness measure for quasi-arithmetic means. IEEE Trans. Fuzzy Syst. **14**(6), 837–848 (2006)
20. Marichal, J.-L.: Tolerant or intolerant character of interacting criteria in aggregation by the Choquet integral. Eur. J. Oper. Res. **155**, 771–791 (2004)
21. Yager, R.R.: Connectives and quantifiers in fuzzy sets. Fuzzy Sets Syst. **40**, 39–76 (1991)
22. Fanaee-T, H., Gama, J.: Event labeling combining ensemble detectors and background knowledge. Prog. Artif. Intell. **2**, 1–15 (2013)

L_1-space for Sugeno Integral

Aoi Honda$^{1(\boxtimes)}$ and Yoshiaki Okazaki^2

^{1} Kyushu Institute of Technology, 680-2 Kawazu, Iizuka, Fukuoka 820-8502, Japan
aoi@ces.kyutech.ac.jp
^{2} Fuzzy Logic Systems Institute, 680-41 Kawazu, Iizuka, Fukuoka 820-0067, Japan
okazaki@flsi.or.jp

Abstract. Let $L_1(Su)$ be the L_1 space with respect to the Sugeno integral for a fuzzy measure [7,9]. $L_1(Su)$ is a linear space with the natural quasi-metric. In general $L_1(Su)$ is not necessarily a topological linear space. We shall characterize explicitly the maximal topological linear subspace of $L_1(Su)$.

Keywords: L_p-space · Dual space · Quasi-metric · Fuzzy measure · Submeasure · Sugeno integral · Translation invariant metric · Quasi-additive functional

1 Introduction

We study the linear topological structure of Sugeno L_1 space $L_1(Su)$ defined by the Sugeno integral for a fuzzy measure. It is well known that the L_1 space for Choquet integral with respect to a submodular fuzzy measure is a normed space [3]. However for Sugeno integral, the L_1 space $L_1(Su)$ is not a normed space. $L_1(Su)$ is a linear space and has a natural translation invariant quasi-metric [7–9]. So that $L_1(Su)$ is an additive topological group, but in genaral it is not a topological linear space. The aim of this paper is characterize the maximal topological linear subspace of $L_1(Su)$ explicitly.

Definition 1. *[4, 5, 7, 9, 10, 14, 16] Let T be a set. A function $\rho(s,t) : T \times T \to [0, +\infty)$ is called a quasi-metric if and only if*

1. $\rho(s,t) \geq 0$, $\rho(s,t) = 0 \iff s = t$, $s,t \in T$
2. $\rho(s,t) = \rho(t,s)$, $s,t \in T$, *and*
3. $\exists K \geq 1$; $\rho(s,t) \leq K\left(\rho(s,u) + \rho(u,t)\right)$, $s,t,u \in T$.

Definition 2. *[2, 3, 11, 13, 15] Let $(X, \mathcal{B}(X))$ be a measurable space on a set X, that is, $\mathcal{B}(X)$ is a σ-algebra on X. A set function $\mu : \mathcal{B}(X) \to [0, +\infty]$ is called a fuzzy measure if and only if*

1. $\mu(\emptyset) = 0$,
2. $A \subset B, A, B \in \mathcal{B}(X) \Rightarrow \mu(A) \leq \mu(B)$.

© Springer International Publishing AG 2017
V. Torra et al. (Eds.): MDAI 2017, LNAI 10571, pp. 63–73, 2017.
DOI: 10.1007/978-3-319-67422-3_7

A fuzzy measure μ is called subadditive (or μ is a submeasure) if and only if

$$\mu(A \cup B) \le \mu(A) + \mu(B)$$

for every $A, B \in \mathcal{B}(X)$.

A fuzzy measure μ is called weakly subadditive if and only if

$$\exists k \ge 1; \ \mu(A \cup B) \le \mu(A) + k\mu(B)$$

for every $A, B \in \mathcal{B}(X)$.

A fuzzy measure μ is said to be continuous from below if and only if

$$\mu(A_n) \uparrow \mu(A)$$

for any $A_n, A \in \mathcal{B}(X)$ such that $A_n \uparrow A$.

A fuzzy measure μ is said to be continuous from above if and only if

$$\mu(B_n) \downarrow \mu(B)$$

for any $B_n, B \in \mathcal{B}(X)$ such that $B_n \downarrow B$ with $\mu(B_1) < +\infty$.

Definition 3. *[11, 15] A set $N \in \mathcal{B}(X)$ is called a strongly null set if and only if*

$$\mu(A \cup N) = \mu(A)$$

for every $A \in \mathcal{B}(X)$.

Lemma 1. *[11, 15] Assume μ is subadditive or weakly subadditive. Then N is a strongly null set if and only if $\mu(N) = 0$.*

Proof. A strongly null set N satisfies $\mu(N) = \mu(\emptyset \cup N) = \mu(\emptyset) = 0$. Conversely assume $\mu(N) = 0$. Then we have for every $A \in \mathcal{B}(X)$, $\mu(A \cup N) \le \mu(A) + k\mu(N) = \mu(A)$ and $\mu(A \cup N) = \mu(A)$ which means N is a strongly null set. $\qquad\square$

A function $f : (X, \mathcal{B}(X)) \to (-\infty, +\infty)$ is called measurable [1,3,6] if for every real number r, it holds that $\{f > r\} := \{x \in X \mid f(x) > r\} \in \mathcal{B}(X)$.

Let $\mu : (X, \mathcal{B}(X)) \to [0, +\infty]$ be a fuzzy measure and $f : (X, \mathcal{B}(X)) \to [0, +\infty)$ be a non-negative measurable function. Then the Sugeno integral [15] of f with respect to μ is defined by

$$(Su) \int_X f d\mu := \sup_{r \ge 0} r \wedge \mu(\{f > r\}).$$

2   Sugeno L_1 Space $L_1(Su)$

For a measurable function $f : (X, \mathcal{B}) \to (-\infty, +\infty)$, we set

$$|f|_1 = \sup_{r \ge 0} r \wedge \mu(|f| > r),$$

$$\mathcal{L}_1 = \{f \mid \ |f|_1 < +\infty\}, \text{ and}$$

$$\mathcal{O}_1 = \{f \in \mathcal{L}_1 \mid \ |f|_1 = 0\}.$$

Lemma 2. *Assume that μ is a weakly subadditive fuzzy measure. Then we have*

$$|f + g|_1 \leq 2k \left(|f|_1 + |g|_1\right), \quad f, g \in \mathcal{L}_1.$$

Proof.

$$|f + g|_1 = \sup_{r \geq 0} r \wedge \mu \left(|f + g| > r\right)$$

$$\leq \sup_{r \geq 0} r \wedge \mu \left(\left\{|f| > \frac{r}{2}\right\} \cup \left\{|g| > \frac{r}{2}\right\}\right)$$

$$\leq \sup_{r \geq 0} r \wedge \left[\mu \left(|f| > \frac{r}{2}\right) + k\mu \left(|g| > \frac{r}{2}\right)\right]$$

$$\leq \left[\sup_{r \geq 0} r \wedge \mu \left(|f| > \frac{r}{2}\right) + \sup_{r \geq 0} r \wedge k\mu \left(|g| > \frac{r}{2}\right)\right]$$

$$= \left[\sup_{r \geq 0} \left(2\frac{r}{2}\right) \wedge \mu \left(|f| > \frac{r}{2}\right) + \sup_{r \geq 0} \left(2\frac{r}{2}\right) \wedge k\mu \left(|g| > \frac{r}{2}\right)\right]$$

$$\leq 2 \sup_{r \geq 0} \left(\frac{r}{2}\right) \wedge \mu \left(|f| > \frac{r}{2}\right) + 2k \sup_{r \geq 0} \left(\frac{r}{2}\right) \wedge \mu \left(|g| > \frac{r}{2}\right)$$

$$- 2|f|_1 + 2k|g|_1 < 2k(|f|_1 + |g|_1),$$

where we have used the inequality $a \wedge (b + c) \leq (a \wedge b + a \wedge c)$. □

Lemma 3. *We have*

$$|cf|_1 \leq \text{Max}\{|c|, 1\}|f|_1 \quad \text{for real number } c \text{ and } f \in \mathcal{L}_1.$$

Proof.

$$|cf|_1 = \sup_{r \geq 0} r \wedge \mu \left(|cf| > r\right)$$

$$= \sup_{r \geq 0} r \wedge \mu \left(|f| > \frac{r}{|c|}\right)$$

$$= \sup_{r \geq 0} \left(|c|\frac{r}{|c|}\right) \wedge \mu \left(|f| > \frac{r}{|c|}\right).$$

If $|c| > 1$, then we have

$$\sup_{r \geq 0} \left(|c|\frac{r}{|c|}\right) \wedge \mu \left(|f| > \frac{r}{|c|}\right) \leq |c| \sup_{r \geq 0} \left(\frac{r}{|c|}\right) \wedge \mu \left(|f| > \frac{r}{|c|}\right) = |c||f|_1.$$

If $|c| \leq 1$, then we have

$$\sup_{r \geq 0} \left(|c|\frac{r}{|c|}\right) \wedge \mu \left(|f| > \frac{r}{|c|}\right) \leq \sup_{r \geq 0} \left(\frac{r}{|c|}\right) \wedge \mu \left(|f| > \frac{r}{|c|}\right) = |f|_1.$$

So that we have the assertion. □

Lemma 4. *Assume that $h \in \mathcal{O}_1$, that is, $|h|_1 = 0$. Then we have*

$$\mu(|h| > r) = 0 \quad \text{for every} \ \ r > 0.$$

In particular, if μ is continuous from below then $h = 0$ μ-almost everywhere, that is

$$\mu(|h| > 0) = 0.$$

Proof. By the definition of $|\ |_1$ we have the assertion. □

Lemma 5. *Let μ be a weakly subadditive fuzzy measure. Then we have*

$$|f \pm h|_1 = |f|_1$$

for every $f \in \mathcal{L}_1$ and $h \in \mathcal{O}_1$.

Proof. By Lemmas 1 and 4, for every $r \geq 0$ it follows that the subset $N(r) := \{|h| > r\}$ is a strongly null set. Let $0 < \varepsilon < 1$ be arbitrarily fixed. Then we have

$$
\begin{aligned}
\mu(|f \pm h| > r) &= \mu\left([\{|f \pm h| > r\} \cap N(\varepsilon r)^c] \cup [\{|f \pm h| > r\} \cap N(\varepsilon r)]\right) \\
&\leq \mu\left([\{|f \pm h| > r\} \cap N(\varepsilon r)^c] \cup N(\varepsilon r)\right) \\
&= \mu(\{|f \pm h| > r\} \cap N(\varepsilon r)^c) \\
&= \mu(|f \pm h| > r, |h| \leq \varepsilon r) \\
&\leq \mu(|f| > (1 - \varepsilon)r).
\end{aligned}
$$

So that we have

$$
\begin{aligned}
r \wedge \mu(|f \pm h| > r) &\leq r \wedge \mu(|f| > (1 - \varepsilon)r) \\
&\leq \frac{1}{(1 - \varepsilon)} \left[(1 - \varepsilon)r \wedge \mu(|f| > (1 - \varepsilon)r)\right] \\
&\leq \frac{1}{(1 - \varepsilon)} |f|_1.
\end{aligned}
$$

Taking $\sup_{r \geq 0}$ in the left hand side, we have

$$|f \pm h|_1 \leq \frac{1}{(1 - \varepsilon)} |f|_1.$$

Letting $\varepsilon \downarrow 0$, we have the assertion. □

Definition 4. *Let μ be a weakly subadditive fuzzy measure. We set*

$$L_1 := \mathcal{L}_1 / \mathcal{O}_1$$

$$\|f + \mathcal{O}_1\|_1 := |f|_1 \quad \text{for} \ f + \mathcal{O}_1 \in L_1.$$

By Lemma 5, the value $\|f + \mathcal{O}_1\|_1$ does not depend on the choice of the representative f of the equivalence class $f + \mathcal{O}_1$. In the sequel we identify the equivalence class $f + \mathcal{O}_1$ with f and write

$$\|f\|_1 = |f + \mathcal{O}_1|_1 \quad \text{for} \ f \in L_1.$$

Then $\|f\|_1$ determines a translation invariant quasi-metric on L_1 as follows.

Theorem 1. *Let μ be a weakly subadditive fuzzy measure. Then the space $(L_1, \|f\|_1)$ is a linear space. The function $\gamma(f, g) := \|f - g\|_1$ is a quasi-metric satisfying :*

1. *$\gamma(cf, 0) \leq \text{Max}\{|c|, 1\}\gamma(f, 0)$ for a real number c and $f \in L_1$,*

2. *$\gamma(f, g) \leq 2k\left(\gamma(f, h) + \gamma(h, g)\right)$ for $f, g, h \in L_1$,*

3. *$\gamma(f + h, g + h) = \gamma(f, g)$ for $f, g, h \in L_1$ (translation invariance of γ).*

Proof. The assertions 1 and 2 follow from Lemmas 3 and 2. The translation invariance is clear. □

Definition 5. *We call the pair $(L_1, \|f\|_1)$ the Sugeno L_1 space and denote it by $L_1(Su)$.*

Remark 1. $L_1(Su)$ is a topological additive group but not necessarily a topological linear space, see Sect. 4.

Remark 2. For $0 < p < \infty$, the Sugeno L_p space $L_p(Su)$ is defined introducing the quasi-metric

$$[f]_p = \left[\sup_{r \geq 0} r \wedge \mu\left(|f|^p > r\right)\right]^{\frac{1}{p}} = \left[\sup_{r \geq 0} r^p \wedge \mu\left(|f| > r\right)\right]^{\frac{1}{p}}.$$

In this case, it holds $L_p(Su) = L_q(Su)$ for every $0 < p, q < \infty$, see [9]

3   Linear Topological Structure of $L_1(Su)$

We shall introduce two function spaces according to [12].

Definition 6. *Let $D \in \mathcal{B}(X)$ and f be a measurable function. Then $\|f\|_{L_\infty(D)} = \|f|_D\|_\infty$, where $f|_D$ is the restriction of f to D.*

Definition 7. $M_\infty = \{f \mid \exists A \in \mathcal{B}(X), \mu(A) + \|f\|_{L_\infty(X \setminus A)} < +\infty\}$

We call M_∞ the truncated L_∞ space. For $f \in M_\infty$, we set

$$\gamma(f) = \inf_{A \in \mathcal{B}(X)} \{\mu(A) + \|f\|_{L_\infty(X \setminus A)}\}.$$

Lemma 6. *Assume μ is weakly subadditive. For $f \in M_\infty$ we have*

$$\gamma(f) = \inf_{r \geq 0} \{r + \mu(|f| > r)\}.$$

Proof. First we show $\gamma(f) \leq \inf_{r \geq 0}\{r + \mu(|f| > r)\}$. If $\inf_{r \geq 0}\{r + \mu(|f| > r)\} = +\infty$, the inequality holds. So that we assume $\ell = \inf_{r \geq 0}\{r + \mu(|f| > r)\} < +\infty$. Then for every $\varepsilon > 0$, there exists $r(\varepsilon)$ such that $r(\varepsilon) + \mu(|f| > r(\varepsilon)) < \ell + \varepsilon$, that is,

$$\mu(|f| > r(\varepsilon)) < \ell + \varepsilon - r(\varepsilon).$$

Let $A = \{|f| > r(\varepsilon)\} \in \mathcal{B}(X)$. Then we have

$$\|f\|_{L_\infty(X \backslash A)} \leq r(\varepsilon).$$

Consequently we have

$$\mu(A) + \|f\|_{L_\infty(X \backslash A)} \leq \ell + \varepsilon.$$

Taking $\inf_{A \in \mathcal{B}(X)}$ and letting $\varepsilon \downarrow 0$, we have the desired inequality.

Conversely we shall prove $\gamma(f) \geq \inf_{r \geq 0}\{r + \mu(|f| > r)\}$. If $\gamma(f) = +\infty$ the inequality is clear. Assume that $m = \gamma(f) < +\infty$. Then for every $\varepsilon > 0$ there exists $A(\varepsilon) \in \mathcal{B}(X)$ such that $\mu(A(\varepsilon)) + \|f\|_{L_\infty(X \backslash A(\varepsilon))} \leq m + \varepsilon$, that is

$$\|f\|_{L_\infty(X \backslash A(\varepsilon))} \leq m + \varepsilon - \mu(A(\varepsilon)).$$

This implies that

$$\mu(x \in X \backslash A(\varepsilon) \mid |f| > m + 2\varepsilon - \mu(A(\varepsilon))) = 0.$$

By the weak subadditivity of μ it follows that

$$
\begin{aligned}
m &+ 2\varepsilon \\
&= m + 2\varepsilon - \mu(A(\varepsilon)) + \mu(A(\varepsilon)) + k\mu(x \in X \backslash A(\varepsilon) \mid |f| > m + 2\varepsilon - \mu(A(\varepsilon))) \\
&\geq m + 2\varepsilon - \mu(A(\varepsilon)) + \mu(A(\varepsilon) \cup \{x \in X \backslash A(\varepsilon) \mid |f| > m + 2\varepsilon - \mu(A(\varepsilon))\}) \\
&\geq m + 2\varepsilon - \mu(A(\varepsilon)) + \mu(\{x \in X \mid |f| > m + 2\varepsilon - \mu(A(\varepsilon))\}).
\end{aligned}
$$

So that we have $\inf_{r \geq 0}\{r + \mu(|f| > r)\} \leq m + 2\varepsilon$. Letting $\varepsilon \downarrow 0$ it follows that $\inf_{r \geq 0}\{r + \mu(|f| > r)\} \leq m = \gamma(f)$. $\qquad\square$

Lemma 7. *Assume that μ is weakly subadditive. Then $\rho(f, g) = \gamma(f - g), f, g \in M_\infty$ is a translation invariant quasi-metric on the truncated L_∞ space M_∞.*

Proof. Let $f, g \in M_\infty$. For every $\varepsilon > 0$, there exist $A, B \in \mathcal{B}(X)$ such that

$$\mu(A) + \|f\|_{L_\infty(X \backslash A)} < \gamma(f) + \varepsilon, \quad \mu(B) + \|g\|_{L_\infty(X \backslash B)} < \gamma(g) + \varepsilon.$$

This implies

$$
\begin{aligned}
\mu(A \cup B) &+ \|f - g\|_{L_\infty(X \backslash A \cup B)} \\
&\leq \mu(A) + k\mu(B) + \|f\|_{L_\infty(X \backslash A)} + \|g\|_{L_\infty(X \backslash B)} \\
&\leq k\left[(\mu(A) + \|f\|_{L_\infty(X \backslash A)}) + (\mu(B) + \|g\|_{L_\infty(X \backslash B)})\right] \\
&< k(\gamma(f) + \gamma(g)) + 2\varepsilon.
\end{aligned}
$$

Letting $\varepsilon \downarrow 0$, it follows that $\gamma(f - g) \leq k(\gamma(f) + \gamma(g))$. $\qquad\square$

Remark 3. If μ is subadditive (k$=$1) then γ is a metric.

Definition 8. *Let M_∞^0 be the linear space of all functions f satisfying the following condition. For every $\varepsilon > 0$, there is $A \in \mathcal{B}(X)$ such that*

$$\mu(A) < \varepsilon, \quad \mu(A) + \|f\|_{L_\infty(X\backslash A)} < +\infty.$$

We call M_∞^0 the finely truncated L_∞ space.

Lemma 8. *If μ is continuous from above then $M_\infty^0 = M_\infty$.*

Proof. Assume $f \in M_\infty$. Let $\varepsilon > 0$ be arbitrarily fixed. There exists $A \in \mathcal{B}(X)$ such that

$$\mu(A) + \|f\|_{L_\infty(X\backslash A)} < +\infty.$$

We set $A_n = \{x \in A \mid |f(x)| > n\}$. Since $A \supset A_n \downarrow \emptyset$ and $\mu(A) < +\infty$, by the continuity from above, there exists N such that $\mu(A_N) < \varepsilon$. Then we have

$$\|f\|_{L_\infty(X\backslash A_N)} \leq \|f\|_{L_\infty(X\backslash A)} + \|f\|_{L_\infty(A\backslash A_N)}$$
$$\leq \|f\|_{L_\infty(X\backslash A)} + N < +\infty.$$

Consequently we have $\mu(A_N) < \varepsilon$ and $\mu(A_N) + \|f\|_{L_\infty(X\backslash A_N)} < +\infty$, which implies $f \in M_\infty^0$. $\qquad\square$

Lemma 9. *Assume μ is weakly subadditive. Then we have $L_\infty \subset M_\infty^0 \subset L_1(Su) = M_\infty$.*

Proof. $L_\infty \subset M_\infty^0$ is clear. We shall show $L_1(Su) = M_\infty$. Assume $f \in M_\infty$. There exists $A \in \mathcal{B}(X)$ such that

$$\mu(A) + \|f\|_{L_\infty(X\backslash A)} < +\infty.$$

Then we have

$$\sup_{r \geq 0} r \wedge \mu(|f| > r)$$

$$= \sup_{0 \leq r \leq \|f\|_{L_\infty(X\backslash A)}} r \wedge \mu(|f| > r) \vee \sup_{r > \|f\|_{L_\infty(X\backslash A)}} r \wedge \mu(|f| > r)$$

$$\leq \|f\|_{L_\infty(X\backslash A)} \vee \mu(|f| > \|f\|_{L_\infty(X\backslash A)}) < +\infty,$$

since

$$\mu(|f| > \|f\|_{L_\infty(X\backslash A)}) \leq \mu(A \cup \{x \in X\backslash A \mid |f(x)| > \|f\|_{L_\infty(X\backslash A)}\})$$
$$\leq \mu(A) + k\mu(\{x \in X\backslash A \mid |f(x)| > \|f\|_{L_\infty(X\backslash A)}\})$$
$$= \mu(A) + 0 = \mu(A) < +\infty.$$

This means $f \in L_1(Su)$ and $M_\infty \subset L_1(Su)$ holds.
Conversely assume $f \in L_1(Su)$, that is,

$$\|f\|_1 = \sup_{r \geq 0} r \wedge \mu(|f| > r) < +\infty.$$

Then there exists r_0 such that $\mu(|f| > r_0) < +\infty$. In fact, if

$$\mu(|f| > r) = +\infty \quad for \ all \ \ r > 0,$$

then $r \wedge \mu(|f| > r) = r \wedge +\infty = r$ and $\|f\|_1 = \sup r = +\infty$, which is a contradiction. If we set $A = \{|f| > r_0\}$, then $\mu(A) < +\infty$ and for every $x \in X \backslash A$, we have $|f(x)| \leq r_0$, which implies $\|f\|_{L_\infty(X \backslash A)} \leq r_0$. It follows that $\mu(A) + \|f\|_{L_\infty(X \backslash A)} < +\infty$ and $f \in M_\infty$. □

Lemma 10. *Assume μ is weakly subadditive. For every $A \in \mathcal{B}(X)$ we have*

$$\|f\|_1 \leq \mu(A) \vee \|f\|_{L_\infty(X \backslash A)} \leq \mu(A) + \|f\|_{L_\infty(X \backslash A)}.$$

Proof.

$$\|f\|_1 = \sup_{r \geq 0} r \wedge \mu(|f| > r)$$

$$= \sup_{0 \leq r < \|f\|_{L_\infty(X \backslash A)}} r \wedge \mu(|f| > r) \vee \sup_{r \geq \|f\|_{L_\infty(X \backslash A)}} r \wedge \mu(|f| > r)$$

$$\leq \|f\|_{L_\infty(X \backslash A)} \vee \mu(A).$$

The second term is obtained by

$$\mu(|f| > r) \leq \mu(A \cup \{x \in X \backslash A \mid |f(x)| > \|f\|_{L_\infty(X \backslash A)}\})$$
$$\leq \mu(A) + k\mu(x \in X \backslash A \mid |f(x)| > \|f\|_{L_\infty(X \backslash A)})$$
$$= \mu(A),$$

where $r \geq \|f\|_{L_\infty(X \backslash A)}$. □

Proposition 1. *Assume μ is weakly subadditive. The identity mapping $(M_\infty, \gamma(f)) \to (L_1(Su), \|f\|_1)$ is continuous.*

Proof. By the preceding Lemma, we have

$$\|f\|_1 \leq \inf_{A \in \mathcal{B}(X)} \{\mu(A) + \|f\|_{L_\infty(X \backslash A)}\} = \gamma(f).$$

□

Proposition 2. *Assume μ is weakly subadditive. Then $(M_\infty^0, \gamma(f))$ is a topological linear space.*

Proof. Assume $t_n \to t_0$ and $f_n \to f_0$, that is $\gamma(f_n - f_0) \to 0$. Then we have

$$\gamma(t_n f_n - t_0 f_0) \leq k \left(\gamma(t_n(f_n - f_0)) + \gamma((t_n - t_0)f_0) \right).$$

Let $K > 0$ be $|t_n| \leq K$ for all n. The first term converges to 0 as follows.

$$\gamma(t_n(f_n - f_0)) = \inf_{A \in \mathcal{B}(X)} \{\mu(A) + \|t_n(f_n - f_0)\|_{L_\infty(X \setminus A)}\}$$
$$\leq \inf_{A \in \mathcal{B}(X)} \{\mu(A) + K\|f_n - f_0\|_{L_\infty(X \setminus A)}\}$$
$$\leq (K \vee 1) \inf_A \{\mu(A) + \|f_n - f_0\|_{L_\infty(X \setminus A)}\}$$
$$= (K \vee 1)\gamma(f_n - f_0) \to 0.$$

To prove the second term $\to 0$, let $\varepsilon > 0$ be arbitrarily fixed. Then there exists $A = A(\varepsilon)$ such that

$$\mu(A) < \varepsilon, \quad \|f_0\|_{L_\infty(X \setminus A)} < +\infty.$$

Then we have

$$\gamma((t_n - t_0)f_0) \leq \mu(A) + \|(t_n - t_0)f_0\|_{L_\infty(X \setminus A)}$$
$$= \mu(A) + |t_n - t_0|\|f_0\|_{L_\infty(X \setminus A)}.$$

So that we have $\limsup \gamma((t_n - t_0)f_0) \leq \mu(A) < \varepsilon$, which implies $\gamma((t_n - t_0)f_0) \to 0$. □

Theorem 2. *Assume μ is weakly subadditive. Then $(M_\infty^0, \|f\|_1)$ is a topological linear space.*

Proof. Assume $t_n \to t_0$ and $f_n \to f_0$, that is $\|f_n - f_0\|_1 \to 0$. Then we have

$$\|t_n f_n - t_0 f_0\|_1 \leq 2k \left(\|t_n(f_n - f_0)\|_1 + \|(t_n - t_0)f_0\|_1\right).$$

Let $K > 0$ be $|t_n| \leq K$ for all n. The first term converges to 0 as follows.

$$\|t_n(f_n - f_0)\|_1 = \sup_{r \geq 0} r \wedge \mu(|t_n(f_n - f_0| > r)$$
$$\leq \sup_{r \geq 0} r \wedge \mu(K|(f_n - f_0| > r)$$
$$= \sup_{s \geq 0}(Ks) \wedge \mu(|(f_n - f_0| > s)$$
$$\leq (K \vee 1)\|f_n - f_0\|_1 \to 0.$$

To prove the second term $\to 0$, let $\varepsilon > 0$ be arbitrarily fixed. Then there exists $A = A(\varepsilon)$ such that

$$\mu(A) < \varepsilon, \quad \|f_0\|_{L_\infty(X \setminus A)} < +\infty.$$

Then we have by Lemma 10 and by Proposition 1

$$\|(t_n - t_0)f_0\|_1 \leq \mu(A) + \|(t_n - t_0)f_0\|_{L_\infty(X \setminus A)}$$
$$= \mu(A) + |t_n - t_0|\|f_0\|_{L_\infty(X \setminus A)}.$$

So that we have $\limsup \|(t_n - t_0)f_0\|_1 \leq \mu(A) < \varepsilon$, which implies $\|(t_n - t_0)f_0\|_1 \to 0$. □

Theorem 3. *Assume μ is weakly subadditive. Then $(M_\infty^0, \|f\|_1)$ is the maximal topological linear subspace of $L_1(Su)$.*

Proof. Let Q be any topological linear subspace of $L_1(Su)$. Then for every $f \in Q$, it holds that

$$t_n \cdot f \to 0 \quad \text{for any sequence } t_n \to 0.$$

Since

$$\|t_n \cdot f\|_1 = \sup_{r \geq 0} r \wedge \mu(|t_n \cdot f| > r) \geq s \wedge \mu(|t_n \cdot f| > s) \to 0 (n \to \infty),$$

we have $\mu(|t_n \cdot f| > s) \to 0 (n \to \infty)$ for every $s \geq 0$ and for every $t_n \to 0$. Consequently it follows that

$$\lim_{R \to \infty} \mu(|f| > R) = 0.$$

For any $\varepsilon > 0$, there exists $R = R_\varepsilon > 0$ such that $\mu(|f| > R) < \varepsilon$. Let $A = \{|f| > R\}$, then we have

$$\mu(A) < \varepsilon, \quad \|f\|_{L_\infty(X \setminus A)} \leq R < +\infty,$$

which shows $f \in M_\infty^0$. So we have $Q \subset M_\infty^0$. $\qquad \square$

Corollary 1. $M_\infty^0 = \{f \in L_1(Su) \mid \lim_{R \to \infty} \mu(|f| > R) = 0\}$.

Corollary 2. *Assume μ is weakly subadditive, $\mu(X) < +\infty$ and μ is continuous from above. Then $(L_1(Su), \|f\|_1)$ is a topological linear space.*

4   Example

Let μ be a fuzzy measure on $[0, \infty)$ defined by

$$\mu(A) = \begin{cases} 1, & \sup A = +\infty, \\ 0, & \text{otherwise.} \end{cases}$$

Then μ is subadditive. For the function $f(x) = x$,

$$\|tf\|_1 = \sup_{r \geq 0} r \wedge \mu(|tf| > r) = \sup_{r \geq 0} r \wedge 1 = 1.$$

This shows the scalar multiplication is not continuous and $(L_1(Su), \|f\|_1)$ is not a topological linear space.

5   Concluding Remarks

The function spaces $L_p(0 < p < \infty)$ are defined for fuzzy integrals other than Choquet or Sugeno integrals [8,9]. These function spaces are not necessarily normed spaces nor metric spaces. However these function spaces admit natural translation invariant quasi-metric structures in many cases. The linear topological properties of these L_p spaces are not widely investigated. In the case where $p = 0$ and $p = \infty$, L_0 is a metric space and L_∞ is a normed space as in the case of the additive measure.

Acknowledgments. This work was supported by JSPS KAKENHI Grant number 26400155, 15K05003

References

1. de Barra, G.: Introduction to Measure Theory. Van Nostrand Reinhold Company, New York (1974)
2. Choquet, G.: Theory of capacities. Ann. Inst. Fourier **5**, 131–295 (1953)
3. Denneberg, D.: Non-additive Measure and Integral. Theory and Decision Library B, vol. 27. Springer, Dordrecht (1994)
4. Frink, A.H.: Distance-functions and the metrization problem. Bull. Am. Math. Soc. **43**, 133–142 (1937)
5. Heinonen, J.: Lectures on Analysis on Metric Spaces. Universitext, Springer (2001)
6. Hewitt, E., Stromberg, K.: Real and Abstract Analysis. Springer, Heidelberg (1965)
7. Honda, A., Okazaki, Y.: Quasi-metric on L_p-space for a fuzzy measure. In: 20th Workshop for Heart and Mind. Japan Society for Fuzzy Theory and Intelligent Informatics (SOFT) (2015)
8. Honda, A., Okazaki, Y.: L_p-space and its dual for a fuzzy measure. In: The 13th International Conference on Modelling Decisions for Artificial Intelligence (MDAI 2016), Andorra, 19–21 September 2016
9. Honda, A., Okazaki, Y.: Quasi-metrics on the function spaces $L_p(0 < p < \infty)$ for a fuzzy measure. In: RIMS Conference on the Structure of Function Spaces and its Related Fields, RIMS Kokyuroku (2017, to appear)
10. Honda, A., Okazaki, Y., Sato, H.: Approximations and the linearity of the Shepp space. Kyushu J. Math. **69**(1), 173–194 (2015)
11. Murofushi, T., Sugeno, M.: A theory of fuzzy measures: representations, the choquet integral and null sets. JMAA **159**, 534–549 (1991)
12. Okazaki, Y.: Topological linear subspace of $L_0(\Omega, \mu)$ for the infinite measure μ. Nihonkai Math. J. **27**, 145–154 (2016)
13. Pap, E.: Null-Additive Set Functions. Kluwer Academic Publishers, Dordorechet (1995)
14. Schroeder, V.: Quasi-metric and metric spaces, arXiv:math/0607304v1 [math:MG], 13 July 2006
15. Sugeno, M., Murofushi, T.: Fuzzy Measure, Lectures on Fuzzy Measure, vol. 3. Nikkan Kogyo Shinbun LTD (1991). (in Japanese)
16. Triebel, H.: A new approach to function spaces on quasi-metric spaces. Revista Matematica Complutense **18**(1), 7–48 (2005)

Clustering and Classification

An Adaptation of the ML-kNN Algorithm to Predict the Number of Classes in Hierarchical Multi-label Classification

Thissiany Beatriz Almeida[(✉)] and Helyane Bronoski Borges

Federal University of Technology – Paraná, Ponta Grossa, Brazil
thissianyalmeida@alunos.utfpr.edu.br,
helyane@utfpr.edu.br

Abstract. The classification problems described in the Machine Learning literature usually relate to the classification of data in which each example is associated to a class belonging to a finite set of classes, all at the same level. However, there are classification issues, of a hierarchical nature, where the classes can be either subclasses or super classes of other classes. In many hierarchical problems, one or more examples may be associated with more than one class simultaneously. These problems are known as hierarchical multi-label classification (HMC) problems. In this work, the ML-KNN algorithm was used to predict hierarchical multi-label problems, in order to determine the number of classes that can be assigned to an example. Through the experiments performed on 10 protein function databases and the statistical analysis of the results, it can be shown that the adaptations performed in the ML-KNN algorithm brought significant performance improvements based on the hierarchical precision and recall metrics Hierarchical.

Keywords: HMC · ML-KNN · Machine learning

1 Introduction

Data mining has become a powerful tool for decision-making, aiding in the retrieval of information, since manual classification or use of simple computational methods became impracticable [13].

As an example of use, one can highlight the categorization of texts and images [1, 2, 12], protein prediction [3, 4], classification of musical genres [5], diagnosis and treatment of diseases and development of drugs through uncovered knowledge [15].

The process of data classification in Machine Learning aims to assign a class to a new example from its characteristics (attributes). Classification problems can be divided according to the dependency relationship among classes in Flat Classification and Hierarchical Classification [14].

The ML classification task can also be categorized according to the number of classes to be estimated for a given example. Thus, this categorization can be applied to traditional problems (single-label) or multi-label problems.

© Springer International Publishing AG 2017
V. Torra et al. (Eds.): MDAI 2017, LNAI 10571, pp. 77–88, 2017.
DOI: 10.1007/978-3-319-67422-3_8

The initial motivation for the searches in the area of multi-label classification arose with the difficulty caused by ambiguities in problems of texts categorization [12]. The hierarchical multi-label classification is considered a relatively new research area [3, 10, 17], which provides the interest of researchers from different areas. The hierarchy of the classes in Problems of Hierarchical Multi-Label Classification can be represented through Trees or Directional Acyclic Graphs, being the last one presents greater complexity because a class can be the daughter of more than one class, unlike the tree.

The purpose of this work is to study the Problems of Hierarchical Multi-Label Classification, in which the relationship between classes is represented in Directed Acyclic Graph (DAG). Consideration is given to a hierarchical organization with the aim of increasing predictive capacity.

In this work, the ML-KNN algorithm is used to determine the number of classes to be assigned to a real example that does not belong to the training base. The main motivation of this work is related to the reduced number of classification algorithms found in the literature for the type of problem addressed, due to being more complex because one must preserve the class hierarchy.

The next sections are organized as follows: Sect. 2 presents the concepts of multi-label hierarchical classification, Sect. 3 addresses the evaluation metrics for hierarchical classification, Sect. 4 presents the statistical test used, Sect. 5 presents the methodology of in this work, Sect. 6 describes the experiments performed with databases of the functional genomic area and discussion about the results obtained with the variation of threshold values and k-neighbors and Sect. 7 the conclusions of the work done.

2   Hierarchical Multi-label Classification

Hierarchical classification problems aim to classify each new input data into one of the leaf nodes providing a more specific and useful knowledge [6]. It may occur, however, that the classifier does not have the desired reliability in the classification in one of the deeper level classes, and it is safer to perform the classification at the higher levels.

The multi-label hierarchical classification has emerged as a new category of classification problems, with characteristics of both multi-label classification problems and hierarchical classification problems. Problems belonging to this new category are called multi-label hierarchical classification problems (HMC).

In a hierarchical multi-label classification problem, an example can belong to multiple classes at the same time and these classes are organized in a hierarchical manner. The hierarchy can be represented in the tree format or a Directed Acyclic Graph (DAG). Thus, an example belonging to a class automatically belongs to all its super classes.

The main difference between the tree structure and the DAG structure is that, in the tree structure, each node except the root node has only one parent node, whereas in the DAG each node, except the root node, may have one or more parent nodes.

Several methods can be used in the treatment of hierarchical multi-label classification tasks. In the literature, there are several papers proposing and analyzing approaches and methods for treatments of multi-label hierarchical problems (HMC) [3, 10, 14, 17], however, there is no consensus on which algorithm to use for the treatment of hierarchical problems Multi-label.

It can also be said that problems of multi-label hierarchical classification are more complex than the other classification problems, since the classes involved in the problem, besides being structured in a hierarchy, the examples can belong to more than one class at the same time.

3   Methods of Evaluation for Hierarchical Classification

The evaluation measures commonly used in data classification do not take into account that complexity rises according to the depth of the class hierarchy. Based on this, in [14] two evaluation measures were proposed based on conventional measures of precision and recall, which take into account hierarchical relationships between classes. These measures were called hierarchical precision and hierarchical recall, take into account classifications in the internal nodes, and leaf nodes.

Each example belongs not only to its class but also to all ancestors of that class in the hierarchical structure. In this way, given any example, with the set of examples, the set of classes predicted for the example, and the set of true classes of the example, sets and can be understood to contain their corresponding ancestor classes as follows:

The hierarchical precision and the hierarchical recall are calculated according to Eqs. (1) and (2), respectively

$$Prec = \frac{\sum_i |\widehat{Y_i'} \cap \widehat{Y_i}|}{\sum_i \widehat{Y_i}} \tag{1}$$

$$Rev = \frac{\sum_i |\widehat{Y_i'} \cap \widehat{Y_i}|}{\sum_i \widehat{Y_i'}} \tag{2}$$

These measures count the number of correctly predicted classes along with the number of ancestor classes of these correctly predicted classes, assuming that examples also belong to the ancestors of their correct classes [16].

The hierarchical precision and recall used alone are not sufficient for the evaluation of classifiers. Therefore, the Prec and Rev Measures must be combined in a hierarchical extension of the F-Measure measure, called FM, presented in Eq. (3). In equation β, it refers to the importance attributed to the values of Prec and Rev. when the value of β is increased, the weight attributed to the Rev value is increased, and when β is decreased, the weight assigned to the value of Prec.

$$FM = \frac{(\beta^2 + 1) * Prec * Rev}{\beta^2 * Prec + Rev} \tag{3}$$

Weights commonly used for β are: 1 (same weights to the accuracy and recall), 2 (recall is double the accuracy) and 0.5 (accuracy is double the recall). The precision has a greater weight for values of $\beta < 1$, whereas $\beta > 1$ favors the recall [17].

4   Statistical Tests

Statistical analysis of the results obtained from a given study is a very important tool in validating this data. Statistical tests are fundamentally used in research that aims to compare experimental conditions.

There are a number of statistical tests that can aid in research and these tests can be divided into parametric and non-parametric. The difference between both tests refers to the type of values of the studied variable. In parametric tests, the values of the variable must have normal distribution or normal approximation. Non-parametric tests, however, do not have any requirements regarding the knowledge of the distribution of the variable in the population.

Due to the lack of knowledge of the distribution of data, experiments performed to deal with multi-label hierarchical problems are used non-parametric tests and the one was chosen for this study was the Wilcoxon Test.

The Wilcoxon test (1945) is applied when two related groups are compared and the variable must be ordinal measurement. The test ranks the difference between the algorithms on each base used for performance evaluation. Then, add the positive differences presented in Eq. (4) and the negative ones in Eq. (5).

$$W^+ = \sum_{d_i > 0} r_i \tag{4}$$

$$W^- = \sum_{d_i < 0} r_i \tag{5}$$

Where r_i is rank of the i-th base evaluated considering the differences between the algorithms compared.

Then, the value T is calculated in Eq. (6), which represents the smallest of the sums of the same signal stations [8].

$$T = \min\left(W^+; W^-\right) \tag{6}$$

Then the value of N, which is the total of the signal differences, is determined. If $N \leq 25$, the critical values of T are tabulated, where N represents the database number evaluated, discounting the number of draws ($d_i = 0$). For the values where $N > 25$ the z statistic is used, in Eq. (7), since the distribution of the data is considered approximately normal.

$$z = \frac{T - \frac{1}{4}n(n+1)}{\sqrt{\frac{1}{24}n(n+1)(2n+1)}} \tag{7}$$

The null hypothesis assumes that the performance difference between algorithms is not significant. With confidence level $\alpha = 0.05$, the null hypothesis cannot be rejected if $-1.96 \leq z \leq 1.96$.

5 Methodology

This work aims to adapt a technique used in AI that applied to hierarchical problems multi-label determines the number of classes that can be assigned to a real example not belonging to the training base. For this, the methodology used in the work can be divided into three phases: a study of the fundamental concepts of hierarchical multi-label classification and its methods, application of the methods in different databases, evaluation, and analysis of the results.

5.1 Dataset

The data set used in the experiments are biological data of the functional genomic area (GO – Gene Ontology), as below (Table 1):

Table 1. Basic properties of the datasets GO

Dataset	#Instances	#Attributes	#Classes	#Levels
Cellcycle	3751	77	4125	13
Church	3749	27	4125	13
Derisi	3719	63	4119	13
Eisen	2418	79	3573	13
Expr	3773	551	4131	13
Gasch1	3758	173	4125	13
Gasch2	3773	52	4131	13
Pheno	1586	69	3127	13
Seq	3900	478	4133	13
Spo	3697	80	4119	13

5.2 Tools

The Mulan framework [14] was used to conduct the experiments together with the ML-kNN classification method. Mulan works in conjunction with Weka Java classes, an environment known and used by the machine learning and data mining community [9]. The package contains a framework that implements a wide range of multi-label methods, which can be used and extended, as well as various evaluation measures.

The format required by Mulan involves two files, one of which is an ARFF (Attribute-Relation File Format) type and the other is an XML (eXtensible Markup Language) type.

In the ARFF file, you can define the type of data being loaded and then provide your own data. In the file, each column was defined and what each column contains, we provide each row of data in a comma-delimited format.

The ARFF file adopted by Mulan has small differences with respect to the ARFF files used in other tools. One difference is that unlike other tools, Mulan does not use only a hierarchical type attribute for the class by defining the hierarchy shortly after this attribute. In Mulan, each label turns a class attribute of type {0, 1}, where 1 represents that example belongs to class and 0 is the absence of that class.

The XML file is responsible for representing the hierarchy/dependency between the labels/classes of the base to be parsed. This file is necessary during the classification and evaluation steps.

It is important to note that this file can only be used to represent multi-label hierarchical problems that have the hierarchy structured in the type of tree, which are simpler structures, where a child node has only one parent node.

The bases chosen for this work have their hierarchy structured in DAG-type format, being this structure more complex when compared to the tree type, since a child node can have multiple parents. Due to this fact, it was necessary to create a new way to represent the hierarchy of the labels, thus replacing the need to provide the XML type file as input, and allowing the use of any hierarchical structure of the DAG-type.

All changes made in this framework will be submitted to the development team for possible incorporation and will be available to the researchers of the area in the tool's own site.

5.3   Methodology of the Experiments

The methodology used to perform the experiments is formed by two main steps: preprocessing and methods, as below:

1. Pre-processing of databases: After obtaining the bases, an algorithm was developed that has as main objective to make the adaptation of ARFF files that are used by the Mulan framework. This algorithm is composed of the following phases:
   a. Input: The algorithm receives two text files. The first file contains the descent ancestry relationship between the labels, and the second contains the values of each attribute of the examples contained in the database and the associated classes associated with each example.
   b. Processing: The first file that is received by the algorithm, allows storing in a list all the labels in the base so that later, they are consulted in the writing of the output file. As mentioned previously, Mulan does not only have a class attribute of the hierarchical type and this has an impact on the assigned value because in the other files the classes are represented in the same attribute and separated by the "@" character. For this, the developed algorithm reads the second input file, verifies instance by instance, which class it belongs to and defines with value 1 the attribute that represents that class, being that the others receive value 0.
   c. Output: As an output, the training and test database files are obtained in the format required by the Mulan framework.
2. The technique used: In order to perform the task of predicting the number of classes in the classification of structured hierarchical data, mainly in the form of a DAG, some changes were made in the ML-kNN algorithm. Since none of the works found were based in DAG format, only in a special type of DAG, this structure is called tree.

ML-kNN is an algorithm known as lazy because it does not learn a compact data model, it just memorizes the training set examples in memory. To classify a new instance of the test base, this instance is compared to all instances belonging to the training set by means of distance calculation.

It has a more complex operation than the traditional kNN algorithm, because, in multi-label problems, the examples have a set of labels and not just a class more popular among the k-neighbors with less value of distance, as it happens in the problems unlabeled.

This algorithm determines the set of labels of the sample to be classified, based on the maximum a posteriori probability calculated from the frequency of each label among the nearest neighboring k, compared to the frequency in all examples.

In multi-label problems, there is another difference, which is related to the evaluation of class prediction, because in single-label problems, the prediction is correct or wrong, and in multi-label problems due to label dependence, the prediction may be partially correct, where some Labels are predicted correctly, but others are not.

As previously mentioned, the ML-kNN algorithm is implemented in the Mulan tool, but it has some barriers to multi-label hierarchical classification. In order to overcome these barriers, the strategy of creating a separate Mulan framework project was adopted, in which only classes related to the ML-kNN classifier would be added, making development easier and more practical, but still respecting the organization of the framework. This new project may be incorporated in the future.

The first change was to create a new input file format to represent the hierarchy of the labels since the Mulan XML file was only able to represent structured bases in the tree-like format. It was then chosen to receive as input to the algorithm a text file that contained the dependency information between the labels.

Through this file, it is possible to assemble the structure of a directed acyclic graph, better known as DAG, because it is possible to store the list with the name of the vertices and their k-neighbors. This file preserves the dependence between the labels, being this an important factor both for the classification and for the accomplishment of the calculations in the evaluation of the performance of the algorithm.

The ARFF file required by the Mulan is still maintained, and for the realization of the experiments, the bases generated in the preprocessing step are used. When granting the input files, the next change was to create new methods in the classes imported from Mulan, these being MLkNN, MultiLabelKNN, MultiLabelLearner, Multi-LabelLearnerBase, and MultiLabelOutput.

The creation of new methods was necessary because of the modification of the structure that the XML file represented and was stored, being now represented by a text file and stored in a graph.

The reading of the base files remains the same, using the Weka library to interpret the ARFF format files. After reading the bases, the ML-kNN classifier is instantiated, passing as parameter the value of k-neighbors and the standard smoothing factor for the classification.

For the construction of the training stage, the training base is passed to the classifier and after this process, the prediction is performed, being performed as previously mentioned, each new instance is calculated the Euclidean distance of it with each instance of the base of training.

For the prediction some changes were made, the first one was to establish a value for the threshold. To improve the algorithm's performance concerning to prediction, the technique of Spyromitros et al. [11]. In this work, the technique previously used in the BR-kNN transformation algorithm will be adapted to the ML-kNN algorithm.

This technique consists in considering a confidence value of each label to decide which will be part of the predicted multi-label, namely: the average number of occurrences of the labels in the examples.

Based on this principle, the selection of the labels is done by comparing the value of the confidence of the label with the value of the threshold, if the confidence is higher than the stipulated threshold, the label is considered part of the predicted multilabel. However, this label selection procedure can predict empty multi-labels. In order to cancel this possibility, if no label is selected, the one with the highest trust is considered part of the multi-label by ML-kNN.

After the prediction of the classes, a search in the graph that represents the hierarchy of the classes is carried out, with the purpose of assigning value 1 to the ancestral attributes to the predicted node. This same process of assigning the value 1 to the ancestor attributes is performed with the attributes class of the test database.

In order to evaluate the performance of the ML-kNN algorithm, precision and recall measurements were implemented, in addition to the variation of the k value and threshold.

6 Experiments and Results

The experiments and results were obtained using the ML-kNN classifier. The experiments were carried out using ten databases of the genomic functional area Gene Ontology of public domain, these being structured in DAG format. For the evaluation, variations in the ML-kNN algorithm were used for the values of k and threshold, where k receives the values 3, 5 and 7, while the threshold varies between 0.5, 0.7 and 0.8.

6.1 Experimental Setup

To perform the experiments were used computers with the Windows 7 Operating System, core i5 processor, 4 Gb of RAM and 1 Tb of HD.

In order to illustrate the proposed modifications, the datasets used in the experiments are biological data from the functional genomic area (GO - Gene Ontology).

The ML-kNN was performed by varying the parameter k, which represents the value of the nearest k-neighbors, assuming the values of 3, 5 and 7. The threshold value was also varied, this being the cut-off value used in comparison with the value of each label of a test instance, between 0.5, 0.7 and 0.8. For the performance evaluation, precision, recall, and F-Measure measurements were used, these being already described previously.

As previously mentioned, the experiments use the classes implemented in the Mulan framework. The holdout strategy is used, where the database is divided into 2/3 for the training base and 1/3 for the test base. The graphs were generated using the Excel software and the Wilcoxon statistical tests generated by an applet using the Two Paired Sample Signed Rank Test method.

6.2 Experimental Setup

The following is the experimental results obtained after the application of the classification algorithm in a multi-label hierarchical database.

The evaluation measures were calculated for each base by performing all possible combinations of k and threshold values, in order to observe the behavior of the prediction process of the number of classes, since the k and threshold values directly influence the process.

These hierarchical measures take into account not only the predicted node but also all of its ancestors, since a prediction in multi-label problems may be correct, wrong, or partially correct. The measures applied were Hierarchical Precision, Hierarchical Revocation, and Hierarchical F-Measure.

After the calculation of the precision and recall measures, the F-Measure measurement is calculated, which is the harmonic measure of the other two measurements. For this, we used the Excel tool to generate the values of all bases. For better visualization, these measurements were plotted in graphs according to Figs. 1, 2 and 3.

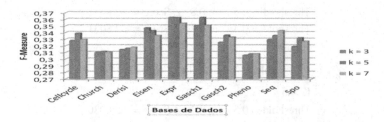

Fig. 1. Metric F-Measure with threshold 0.5

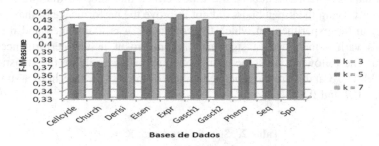

Fig. 2. Metric F-Measure with threshold 0.7

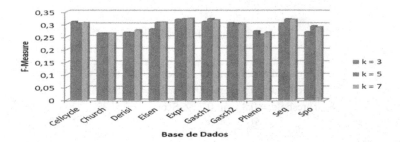

Fig. 3. Metric F-Measure with threshold 0.8

Through the graphs, it is possible to notice that in the same base, the variation of the measurement of F-Measure is little significant. In addition, that its value is close to 0, demonstrating that when the measures of precision and recall are compared, the algorithm is not very effective.

6.3   Statistical Evaluation

To perform the Wilcoxon statistical test, the values of the precision measurement were chosen. To compare the results, the threshold and k-neighbors variants were used. In the first moment, the constant k was maintained and the threshold was varied and in the second moment the constant threshold was maintained and the value of kneighbors was varied.

For the two variations performed from the Wilcoxon test, the null hypothesis assumes that the performance difference between algorithms is not significant. With confidence level, $\alpha = 0.05$ and the null hypothesis cannot be rejected if $-1.96 \leq z \leq 1.96$.

After realizing the experiments, it can be verified that keeping the threshold value constant by varying the k-neighbors did not improve the performance of the algorithm. However, in the experiment in which the constant k was maintained and the cut-off value was varied, there was a significant improvement in the performance of the algorithm. The following are some results obtained when applying the statistical test.

In Tables 2, 3 and 4 are the tabulated values of the test for threshold assuming the values 0.5, 0.7 and 0.8.

Table 2. Statistical test with K = 3

K = 3	
Threshold = 0.5 com 0.7	Z-Score = −2.8031
Threshold = 0.5 com 0.8	Z-Score = −2.8031
Threshold = 0.8 com 0.7	Z-Score = −2.8031

Table 3. Statistical test with K = 5

K = 5	
Threshold = 0.5 com 0.7	Z-Score = −2.8030
Threshold = 0.5 com 0.8	Z-Score = −2.8030
Threshold = 0.8 com 0.7	Z-Score = −2.8030

Table 4. Statistical test with K = 7

K = 7	
Threshold = 0.5 com 0.7	Z-Score = −2.8032
Threshold = 0.5 com 0.8	Z-Score = −2.8032
Threshold = 0.8 com 0.7	Z-Score = −2.8032

Instead, to the result obtained in the variation of the values of k, it is possible to notice that in this test, the obtained values of, are superior to the limit of -1.96, which means that the null hypothesis can be rejected and that there are improvements in the performance of the algorithm.

7 Experiments and Results

In order to carry out the work, a methodology was created that includes the following phases: Pre-Processing of databases, Adaptation of ML-kNN Hierarchical Classification algorithm, Application of algorithm in databases and Statistical evaluation of results.

During the development of the work, some changes were necessary to use the MLkNN multi-label classification algorithm. Among these changes, we can mention the modification of the input file (XML format) of the Mulan framework, where the algorithm is implemented. We can also emphasize the fact that the technique used previously only in the BR-kNN algorithm if in the prediction phase no class is assigned to the instance, it assigns to that instance the class that has the highest confidence value.

Through the experiments performed using the adaptation of the ML-kNN algorithm, the relationship between the cut-off variable and the precision measure was verified. When performing the variation of the threshold, it was observed that the adapted algorithm has a superior performance when this variable receives the value of 0.8, being this fact explained in the comparison between the confidence value of the labels with the threshold, because the higher the value of the Threshold, the tendency is for a smaller number of classes to be predicted, entering the condition where only the class with the highest confidence among the labels of that instance is assigned. From this, it is known that the number of classes of intercession between calculates precision true and predicted divided by the number of predicted classes, as in this case, the number of predicates is 1, the tendency is that the accuracy is greater.

References

1. Dumais, S., Chen, H.: Hierarchical classification of web content. In: Proceedings of the 23rd Annual International ACM SIGIR Conference on Research and Development in Information Retrieval, Athens, Greece, pp. 256–263 (2000)
2. Sun, A., Lim, E.-P.: Hierarchical text classification and evaluation. In: Proceedings of the 2001 IEEE International Conference on Data Mining. IEEE Computer Society, pp. 521–528 (2001)
3. Costa, E.P., Lorena, A.C., Carvalho, A.P.L.F., Freitas, A.A.: A review of performance evaluation measures for hierarchical classifiers. In: Proceedings of the AAAI07 - Workshop on Evaluation Methods for Machine Learning II, pp. 1–6 (2007)
4. Holden, N., Freitas, A.: A hierarchical classification of protein function with ensembles of rules and particle swarm optimization. Soft. Comput. **13**, 259–272 (2008)
5. Barutcuoglu, Z., DeCoro, C.: Hierarchical shape classification using Bayesian aggregation. In: Proceedings of the IEEE International Conference on Shape Modeling and Applications, Matsushima, Japan, pp. 44–44 (2006)

6. Carvalho, A.C.P.F., Freitas, A.: A Tutorial on Hierarchical Classification with Applications in Bioinformatics, vol. 1. Idea Group, São Paulo (2007)
7. Cerri, R., Carvalho, A.C.P.L.F., e Costa, E.P.: Classificação hierárquica de proteínas utilizando técnicas de aprendizado de máquina. In: II Workshop on Computational Intelligence, páginas 1–6, Salvador (2008)
8. Guyon, I., Elisseeff, A.: An introduction to feature extraction. In: Guyon, I., Nikravesh, M., Gunn, S., Zadeh, L.A. (eds.) Feature Extraction, Foundations and Applications, vol. 207, pp. 1–24. Springer, Heidelberg (2006)
9. Yang, Y., Pedersen, J.O.: A comparative study on feature selection in text categorization. In: Proceedings of the Fourteenth International Conference on Machine Learning. Morgan Kaufmann Publishers Inc., pp. 412–420 (1997)
10. Spyromitros, E., Tsoumakas, G., Vlahavas, I.: An empirical study of lazy multilabel classification algorithms. In: Hellenic conference on Artificial Intelligence, Berlin, Alemanha, pp. 401–406 (2009)
11. Borges, H.B., Nievola, J.C.: Multi-label hierarchical classification using a competitive neural network for protein function prediction. In: 2012 International Joint Conference on Neural Networks (IJCNN 2012), Brisbane, Austrália, vol. 1, pp. 1–8. IEEE Press, Piscataway (2012)
12. Tsoumakas, G., Katakis, I., Vlahavas, I.: Mining multi-label data. In: Maimon, O., Rokach, L. (ed.) Data Mining and Knowledge Discovery Handbook, 2nd edn. Springer, Boston (2010)
13. Zhang, M.L., Zhou, Z.H.: Ml-kNN: a lazy learning approach to multi-label learning. Pattern Recogn. 40(7), 2038–2048 (2007)
14. Kiritchenko, S., Matwin, S., Famili, A.F.: Hierarchical text categorization as a tool of associating genes with gene ontology codes. In: Proceedings of the Second European Workshop on Data Mining and Text Mining in Bioinformatics, Pisa, Italia (2004)
15. Wilcoxon, F.: Individual comparisons by ranking methods. Biometrics 1, 80–83 (1945)
16. Stojanova, D., Ceci, M., Malerba, D., Džeroski, S.: Learning hierarchical multi-label classification trees from network data. In: Fürnkranz, J., Hüllermeier, E., Higuchi, T. (eds.) DS 2013. LNCS, vol. 8140, pp. 233–248. Springer, Heidelberg (2013). doi:10.1007/978-3-642-40897-7_16
17. Amati, G., Rijsbergen, C.J.V.: Probabilistic models of information retrieval based on measuring the divergence from randomness. ACM Trans. Inf. Syst. (TOIS) 20(4), 357–389 (2002)

On Fuzzy Clustering for Categorical Multivariate Data Induced by Polya Mixture Models

Yuchi Kanzawa[(⊠)]

Shibaura Institute of Technology, Koto, Tokyo 135-8548, Japan
kanzawa@sic.shibaura-it.ac.jp

Abstract. In this paper, three fuzzy clustering models for categorical multivariate data are proposed based on the Polya mixture model and q-divergence. A conventional fuzzy clustering model for categorical multivariate data is constructed by fuzzifying a multinomial mixture model (MMM) via regularizing Kullback-Leibler (KL) divergence appearing in a pseudo likelihood of an MMM, whereas MMM is extended to a Polya mixture model (PMM) and no fuzzy counterpart to PMM is proposed. The first proposed model is constructed by fuzzifying PMM, by means of regularizing KL-divergence appearing in a pseudo likelihood of the model. The other two models are derived by modifying the first proposed algorithm, which is based on the fact that one of the three fuzzy clustering models for vectorial data is similar to the first proposed model, and that another fuzzy clustering model for vectorial data can connect the other two fuzzy clustering models for vectorial data based on q-divergence. In numerical experiments, the properties of the membership of the proposed methods were observed using an artificial dataset.

1 Introduction

The hard c-means (HCM) clustering algorithm [1] splits datasets into well-separated clusters by minimizing the sum of the squared distances between the data and cluster centers. This concept has been extended to fuzzy clustering. Specifically, Bezdek's algorithm replaces linear membership weights with the power of membership [2], thereby producing what is commonly known as the fuzzy c-means (FCM) algorithm. To distinguish this algorithm from the many variants that have been proposed since, this algorithm is referred to as the Bezdek-type fuzzified FCM (bFCM) in this paper. Another fuzzy approach used for cluster analysis is the regularization of the objective function of HCM. HCM is singular, and an appropriate cluster cannot be obtained using the Lagrangian multiplier method. Therefore, Miyamoto and Mukaidono introduced a regularization term into its objective function as the negative entropy of membership [3], thereby producing entropy-regularized FCM (eFCM). A major drawback to the above clustering algorithms is that these algorithms tend to create clusters of equal sizes. As a result, it is possible for a part of a large cluster to be misclassified as one of a smaller cluster if volumes of clusters are out of balance. To avoid such issues, some approaches that use variables to control cluster sizes

© Springer International Publishing AG 2017
V. Torra et al. (Eds.): MDAI 2017, LNAI 10571, pp. 89–102, 2017.
DOI: 10.1007/978-3-319-67422-3_9

have been proposed [4,5]. Such methods that correspond to bFCM and eFCM are referred to bFCMA and eFCMA in this paper. Furthermore, bFCMA has been generalized in [6], and is referred to as the generalized FCM (gFCM).

Clustering for categorical multivariate data is a promising technique for summarizing co-occurrence information such as purchase history transactions and document-keyword frequencies, which are composed of mutual similarities among objects and items. A multinomial mixture model (MMM) [7] is a probabilistic model for clustering tasks for categorical multivariate data, in which component distributions are given by multinomial distributions. The object assignment and item typicalities are iteratively estimated based on the EM algorithm [8]. Honda et al. [9] proposed the Kullback-Leibler (KL) divergence-regularized fuzzy clustering model for categorical multivariate data (FCCM), referred to as KLFCCM, supported by the KL-divergence-based regularization concept. In the KLFCCM method, MMMs are extended to a fuzzy clustering model, in which the degree of fuzziness of object memberships can be controlled by the KL-divergence-based penalty term. However, Madsen et al. [10] have pointed out that the multinomial model cannot capture well the phenomenon that pairs of an object and an item tend to appear in bursts: if a pair of an object and an item appears once, it is more likely to appear again. Then, Madsen et al. [10] proposed the Polya mixture model (PMM). PMM has the potential to be extended to a fuzzy clustering model similar to the manner in which MMM was extended to KLFCCM, which is the primary motivation for this paper.

Several machine learning algorithms for classification and clustering employ a variety of divergence. The most popular and often used divergences are the inner product induced squared distance and Kullback-Leibler (KL)-divergence. The HCM, bFCM, eFCM, bFCMA, eFCMA, and gFCM algorithms use the inner product induced squared distance to measure the dissimilarities between objects and cluster centers. The eFCMA [5] algorithm uses KL-divergence between memberships and variables to control cluster size for regularization of HCM. Recently, advanced machine learning algorithms have used alternative generalized divergences. In particular, q-divergence [11], related to Tsallis entropy [12], might provide more robust solutions with improved accuracy with respect to outliers and additive noise. A fuzzy clustering method based on Tsallis entropy has been proposed in [13]. Furthermore, although it is not described in [6], gFCM can be interpreted as the regularization of bFCMA with q-divergence between memberships and variables controlling cluster sizes, as shown in this paper. However, the methods in [13] and gFCM apply only to vectorial data. This paper examines the potential that multivariate categorical data can be clustered effectively with q-divergence.

In this paper, we propose three clustering algorithms based on PMM and q-divergence. First, the log-likelihood for PMM is fuzzified by introducing a fuzzification parameter into the KL-divergence, similar to the manner in which KLFCCM was fuzzified from MMM. Then, maximization of the fuzzified log-likelihood leads to the first proposed algorithm, which is referred to as KLFCCMP. The KLFC-CMP algorithm reduces into the PMM with a specified parameter. Next, we show that gFCM [6] can be interpreted as the regularization of bFCMA [4] with q-divergence, and that gFCM reduces not only to bFCMA but also reduces to eFCMA [5] with specified sets of parameters. Furthermore, we show that the maximization problem for the first proposed algorithm is similar to the minimization problem for the eFCMA algorithm. Then, based on the knowledge that eFCMA based on KL-divergence is generalized into gFCM based on q-divergence, and that the first proposed algorithm, KLFCCMP, is based on KL-divergence, an FCCM algorithm based on q-divergence is derived from an optimization problem built by extending KL-divergence in KLFCCMP to q-divergence, referred to as qFC-CMP. The qFCCMP algorithm reduces to KLFCCMP with a specified parameter. Finally, based on the knowledge that gFCM with a specified parameter reduces to bFCMA, another FCCM algorithm is derived from an optimization problem built by specifying a parameter value for qFCCMP, which is referred to as bFCCMP. In numerical experiments, we observe the membership properties of the proposed methods using a fuzzy classification function (FCF) [14].

The remainder of this paper is organized as follows. In Sect. 2, notations and the conventional methods are introduced. Section 3 presents the proposed methods. Section 4 provides some numerical examples. Section 5 presents our concluding remarks.

2 Preliminaries

2.1 Entropy and Divergence

For a discrete probability distribution P, Shannon entropy $H_{\mathsf{Shannon}}(P)$ is defined as

$$H_{\mathsf{Shannon}}(P) = - \sum_i P(i) \log(P(i)). \tag{1}$$

For two discrete probability distributions P and Q, the KL-divergence of Q from P, $D_{\mathsf{KL}}(P||Q)$ is defined as

$$D_{\mathsf{KL}}(P||Q) = \sum_i P(i) \log\left(P(i)/Q(i)\right). \tag{2}$$

Shannon entropy and KL-divergence have been used to derive fuzzy clustering [3, 6, 9] for vectorial, spherical, and multivariate categorical data.

Shannon entropy and KL-divergence have been generalized by using a family of functions called generalized logarithmic functions or q-logarithmic function:

$$\log_q(x) = (x^{1-q} - 1)/(1 - q) \quad \text{(for } x > 0\text{)} \tag{3}$$

as

$$H_{\mathsf{Tsallis}}(P) = - \frac{1}{q-1}\left(\sum_i P(i) - 1\right), \tag{4}$$

$$D_q(P||Q) = \frac{1}{q-1}\left(\sum_i P(i)^q Q(i)^{1-q} - 1\right), \tag{5}$$

referred to as Tsallis entropy [12] and q-divergence [11]. In the limit, as $q \to 1$, Shannon entropy and KL-divergence are recovered. Tsallis entropy has been also used to derive fuzzy clustering [13] only for vectorial data and q-divergence has been implicitly used to derive fuzzy clustering only for vectorial data [6], although that is not indicated in the literature. Tsallis entropy and q-divergence have not been used for fuzzy clustering of multivariate categorical data. That fact is the main motivation for this work.

2.2   Fuzzy Clustering for Vectorial Data

In this subsection, three optimization problems are introduced, which lead to the representative three fuzzy clustering methods for vectorial data. These optimization problems and their relation are the basis for the methods proposed in this study.

Let $X = \{x_k \in \mathbb{R}^p \mid k \in \{1, \cdots, N\}\}$ be a dataset of p-dimensional points. The membership of x_k that belongs to the i-th cluster is denoted by $u_{i,k}$ ($i \in \{1, \cdots, C\}, k \in \{1, \cdots, N\}$) and the set of $u_{i,k}$ is denoted by u, which is also known as the partition matrix, which satisfies the constraint

$$\sum_{i=1}^{C} u_{i,k} = 1, \quad u_{i,k} \in [0,1]. \tag{6}$$

The cluster center set is denoted by $v = \{v_i \mid v_i \in \mathbb{R}^p, i \in \{1, \cdots, C\}\}$. The variable controlling the i-th cluster size is denoted by π_i. The i-th element of vector π is denoted by π_i, and π satisfies the constraint

$$\sum_{i=1}^{C} \pi_i = 1. \tag{7}$$

The methods bFCMA, eFCMA, and gFCM are derived by solving the optimization problems

$$\underset{u,v,\pi}{\text{minimize}} \sum_{i=1}^{C} \sum_{k=1}^{N} (\pi_i)^{1-m} (u_{i,k})^m \|x_k - v_i\|_2^2, \tag{8}$$

$$\underset{u,v,\pi}{\text{minimize}} \sum_{i=1}^{C} \sum_{k=1}^{N} u_{i,k} \|x_k - v_i\|_2^2 + \lambda^{-1} \sum_{i=1}^{C} \sum_{k=1}^{N} u_{i,k} \log \left(\frac{u_{i,k}}{\pi_i} \right), \tag{9}$$

$$\underset{u,v,\pi}{\text{minimize}} \sum_{i=1}^{C} \sum_{k=1}^{N} (\pi_i)^{1-m} (u_{i,k})^m \|x_k - v_i\|_2^2 + \lambda^{-1} \sum_{i=1}^{C} \sum_{k=1}^{N} (\pi_i)^{1-m} (u_{i,k})^m, \tag{10}$$

respectively, subject to Eqs. (6) and (7), where $m > 1$ and $\lambda > 0$ are fuzzification parameters. (These algorithms are omitted for brevity.) The gFCM method with $(m-1, \lambda) \to (+0, \infty)$ reduces to bFCMA. Although not described in [6], a reparametrized gFCM can be interpreted as the regularization of bFCMA with

q-divergence between memberships and variables that control cluster sizes. Furthermore, this reparametrized gFCM reduces to eFCMA with a specified parameter value. Then, this reparametrized gFCM connects bFCMA and eFCMA. These are further described in a later section.

2.3   Conventional Fuzzy Clustering for Categorical Multivariate Data Based on KL-divergence

It is assumed that for datasets $A = \{a_k \mid k \in \{1, \ldots, N\}\}$ and $B = \{b_\ell \mid \ell \in \{1, \ldots, M\}\}$, the co-occurrence information between a_k and b_ℓ, $x_k^{(\ell)}$, is given. X is a matrix whose (k, ℓ)-th element is $x_k^{(\ell)}$. We refer to A and B as the object and item sets, respectively. The membership of object a_k belonging to the i-th cluster is denoted by $u_{i,k}$. The (i, k)-th element of matrix u is denoted by $u_{i,k}$, and u satisfies the constraint given in Eq. (6). The typicality of item b_ℓ belonging to the i-th cluster is denoted by $v_i^{(\ell)}$. The (i, ℓ)-th element of matrix v is denoted by $v_i^{(\ell)}$, and v satisfies the constraint

$$\sum_{\ell=1}^{M} v_i^{(\ell)} = 1. \tag{11}$$

The variable controlling the i-th cluster size is denoted by π_i. The i-th element of vector π is denoted by π_i, and π satisfies the constraint given in Eq. (7). Honda et al. proposed an FCCM algorithm [9] induced by MMMs, referred to as KLFCCM, by solving the following optimization problem:

$$\underset{u,v,\pi}{\text{maximize}} \sum_{i=1}^{C} \sum_{k=1}^{N} \sum_{\ell=1}^{M} u_{i,k} x_k^{(\ell)} \log\left(v_i^{(\ell)}\right) + \lambda^{-1} \sum_{i=1}^{C} \sum_{k=1}^{N} u_{i,k} \log\left(\frac{\pi_i}{u_{i,k}}\right), \tag{12}$$

which is subject to Eqs. (6), (7), and (11), where $\lambda > 0$ is a fuzzification parameter. This optimization problem is derived from the pseudo log-likelihood of MMMss described as

$$\sum_{i=1}^{C} \sum_{k=1}^{N} \sum_{\ell=1}^{M} u_{i,k} x_k^{(\ell)} \log\left(v_i^{(\ell)}\right) + \sum_{i=1}^{C} \sum_{k=1}^{N} u_{i,k} \log\left(\frac{\pi_i}{u_{i,k}}\right), \tag{13}$$

where, in the probabilistic framework, $u_{i,k}$ denotes the posterior probability of the i-th component multinomial distribution given the k-th multinomial object, $v_i^{(\ell)}$ denotes the probability of observing the ℓ-th item from the i-th component multinomial distribution, and π_i denotes the prior probability of the i-th component multinomial distribution. The KLFCCM optimization problem is fuzzified by introducing the fuzzification parameter λ into the KL divergence of the posterior from the prior

$$D_{\mathsf{KL}}(u \parallel \pi) = \sum_{k=1}^{N} \sum_{i=1}^{C} u_{i,k} \log\left(\frac{u_{i,k}}{\pi_i}\right), \tag{14}$$

and the degree of fuzziness of object memberships can be controlled by the KL divergence-based penalty term. The KLFCCM optimization problem recovers the MMM optimization problem with $\lambda \to 1$.

2.4 Polya Mixture Model

In a co-occurrence matrix, if an item occurs once, it is likely that the same item will occur again. This phenomenon is called burstiness. The multinomial distribution was used extensively to model the co-occurrence matrix, but it does not account for burstiness. As an alternative to the co-occurrence matrix, the Polya mixture model is proposed [10].

The Dirichlet distribution is defined for a random vector, $v = (v^{(1)}, \ldots, v^{(M)})$, on a simplex of M dimensions. Elements of a random vector on a simplex sum to one. We interpret v as item occurrence probabilities on M items, such that the Dirichlet distribution models item occurrence probabilities. The density function of the Dirichlet for v is

$$\text{Prob}_D(v; \alpha) = \frac{\Gamma\left(\sum_{\ell=1}^{M} \alpha^{(\ell)}\right)}{\prod_{\ell=1}^{M} \Gamma(\alpha^{(\ell)})} \prod_{\ell=1}^{M} (v^\ell)^{\alpha^{(\ell)}-1}, \tag{15}$$

where $\alpha = (\alpha^{(1)}, \ldots, \alpha^{(M)})$ with $\alpha^{(\ell)} > 0$ is a parameter vector, and $\Gamma()$ is the gamma function. When the random vector v as the parameter of a multinomial is drawn from the Dirichlet mixture distribution, defined as

$$\text{Prob}_{DM}(v; \pi, \alpha) = \sum_{i=1}^{C} \pi_i \text{Prob}_D(v_i; \alpha_i), \tag{16}$$

where $\pi = (\pi_1, \ldots, \pi_C)$ is a weight vector for each component Dirichlet distribution and $\alpha = (\alpha_1^{(1)}, \ldots, \alpha_C^{(M)})$ is a parameter matrix, the compound distribution for objects $x = (x^{(1)}, \ldots, x^{(M)})$ is

$$\text{Prob}_{PM}(x; \pi, \alpha) = \sum_{i=1}^{C} \pi_i \frac{\Gamma\left(\sum_{\ell=1}^{M} \alpha_i^{(\ell)}\right)}{\Gamma\left(\sum_{\ell=1}^{M} \alpha_i^{(\ell)} + x^{(\ell)}\right)} \prod_{\ell=1}^{M} \frac{\Gamma\left(x^{(\ell)} + \alpha_i^{(\ell)}\right)}{\Gamma\left(\alpha_i^{(\ell)}\right)}. \tag{17}$$

Each $x^{(\ell)}$ signifies the occurrence frequency of the ℓ-th item in an object. This distribution is called the Polya mixture distribution. Its set of parameters (π, u, α) is estimated given objects $\{x_k\}_{k=1}^{N}$ by maximizing its pseudo log-likelihood or solving the following optimization problem:

$$\begin{aligned}
\underset{\pi, u, \alpha}{\text{maximize}} \sum_{k=1}^{N} \sum_{i=1}^{C} u_{i,k} \log\left(\frac{\pi_i}{u_{i,k}}\right) + \sum_{k=1}^{N} \sum_{i=1}^{C} u_{i,k} \left(\ell\Gamma\left(\sum_{\ell=1}^{M} \alpha_i^{(\ell)}\right)\right. \\
\left. - \ell\Gamma\left(\sum_{\ell=1}^{M} \alpha_i^{(\ell)} + x_k^{(\ell)}\right) + \sum_{\ell=1}^{M} \left(\ell\Gamma\left(x_k^{(\ell)} + \alpha_i^{(\ell)}\right) - \ell\Gamma\left(\alpha_i^{(\ell)}\right)\right)\right),
\end{aligned} \tag{18}$$

which is subject to Eqs. (6) and (7), where $\ell\Gamma$ is the log-gamma function. Its fuzzy counterpart is currently not defined in the literature. This issue is the motivation for this work.

3   Proposed Method

3.1   Basic Concept

Based on the fact that KLFCCM was derived by fuzzifying MMM, in this paper, we derive an FCCM method by fuzzifying a PMM; the result is referred to as KLFCCMP. We can find the KL-divergence of the posterior $u_{i,k}$ from the prior π_i in the first term of the objective function for PMM given in Eq. (18). Then, the optimization problem of the first proposed method is derived by fuzzifying PMM by introducing a parameter $\lambda > 0$ in the KL-divergence of the posterior from the prior, described as

$$
\underset{u,v,\pi}{\text{maximize}}\, \lambda^{-1} \sum_{k=1}^{N} \sum_{i=1}^{C} u_{i,k} \log\left(\frac{\pi_i}{u_{i,k}}\right) + \sum_{k=1}^{N} \sum_{i=1}^{C} u_{i,k} \left(\ell\Gamma\left(\sum_{\ell=1}^{M} \alpha_i^{(\ell)}\right) \right.
$$

$$
\left. - \ell\Gamma\left(\sum_{\ell=1}^{M} \alpha_i^{(\ell)} + x_k^\ell\right) + \sum_{\ell=1}^{M} \left(\ell\Gamma\left(x_k^\ell + \alpha_i^{(\ell)}\right) - \ell\Gamma\left(\alpha_i^{(\ell)}\right)\right) \right), \tag{19}
$$

which is subject to Eqs. (6) and (7). This derivation is similar in manner to the derivation of KLFCCM by fuzzifying MMM by introducing a parameter $\lambda > 0$ in the KL-divergence of the posterior from the prior. This optimization recovers PMM with $\lambda \to 1$.

Next, as one of the two preparations to derive the second proposed method, we recall that gFCM is derived by regularizing bFCMA using the q-divergence. Rewriting the fuzzification parameter λ in Eq. (10) of the gFCM objective function, as $\tilde{\lambda}(m-1)$ with another parameter $\tilde{\lambda}$, we have

$$
\underset{u,v,\pi}{\text{minimize}} \sum_{i=1}^{C} \sum_{k=1}^{N} (\pi_i)^{1-m} (u_{i,k})^m d_{i,k} + \frac{\tilde{\lambda}^{-1}}{m-1} \left(\sum_{i=1}^{C} \sum_{k=1}^{N} (\pi_i)^{1-m} (u_{i,k})^m - 1 \right), \tag{20}
$$

where the first term is the objective function of bFCMA (Eq. (8)). The second term is the q-divergence between memberships $u_{i,k}$ and variables controlling cluster sizes π_i with the parameter of q-divergence, m, and with the fuzzification parameter, $\tilde{\lambda}$. Therefore, gFCM can be interpreted as further fuzzification of bFCMA by q-divergence. Furthermore, it should be noted that gFCM with $m \to 1$ reduces to eFCMA because the first term reduces to HCM and the second term reduces to KL-divergence between memberships $u_{i,k}$ and variables controlling cluster sizes π_i with the parameter of q-divergence, m, and with the fuzzification parameter, $\tilde{\lambda}$. Therefore, gFCM connects bFCMA and eFCMA.

Next, as another preparation to derive the second proposed method, we show that the maximization problem for the first proposed algorithm, KLFCCMP, is similar to the minimization problem for the eFCMA algorithm. The KLFCCMP maximization problem is equivalently described as the following minimization problem:

$$
\begin{aligned}
\operatorname*{minimize}_{u,v,\pi} \lambda^{-1} & \sum_{k=1}^{N}\sum_{i=1}^{C} u_{i,k}\log\left(\frac{u_{i,k}}{\pi_i}\right) + \sum_{k=1}^{N}\sum_{i=1}^{C} u_{i,k}\left(-\ell\Gamma\left(\sum_{\ell=1}^{M}\alpha_i^{(\ell)}\right)\right. \\
& + \ell\Gamma\left(\sum_{\ell=1}^{M}\alpha_i^{(\ell)} + x_k^{(\ell)}\right) - \sum_{\ell=1}^{M}\left(\ell\Gamma\left(x_k^{(\ell)} + \alpha_i^{(\ell)}\right) + \ell\Gamma\left(\alpha_i^{(\ell)}\right)\right)\Bigg),
\end{aligned}
\tag{21}
$$

which is identical to the eFCMA minimizing problem except that the object-cluster dissimilarity in eFCMA is described as $\|x_k - v_i\|_2^2$, whereas the object-item membership relation in KLFCCMP is described as $-\ell\Gamma(\sum_{\ell=1}^{M}\alpha_i^{(\ell)}) + \ell\Gamma(\sum_{\ell=1}^{M}\alpha_i^{(\ell)} + x_k^{(\ell)}) - \sum_{\ell=1}^{M}(\ell\Gamma(x_k^{(\ell)} + \alpha_i^{(\ell)}) + \ell\Gamma(\alpha_i^{(\ell)}))$.

Similar to the manner in which eFCMA is generalized to gFCM by q-divergence, we propose a generalized optimization problem of KLFCCMP, Eq. (19), by fuzzifying with q-divergence as

$$
\begin{aligned}
\operatorname*{minimize}_{u,v,\pi} \frac{\lambda^{-1}}{m-1} & \left(\sum_{k=1}^{N}\sum_{i=1}^{C}(\pi_i)^{1-m}(u_{i,k})^m - 1\right) + \sum_{k=1}^{N}\sum_{i=1}^{C}(\pi_i)^{1-m}(u_{i,k})^m \\
\times & \left(-\ell\Gamma\left(\sum_{\ell=1}^{M}\alpha_i^{(\ell)}\right) + \ell\Gamma\left(\sum_{\ell=1}^{M}\alpha_i^{(\ell)} + x_k^{(\ell)}\right) - \sum_{\ell=1}^{M}\left(\ell\Gamma\left(x_k^{(\ell)} + \alpha_i^{(\ell)}\right) + \ell\Gamma\left(\alpha_i^{(\ell)}\right)\right)\right),
\end{aligned}
\tag{22}
$$

which is subject to Eqs. (6) and (7). This qFCCMP optimization problem with $m - 1 \to +0$ reduces to the KLFCCMP optimization problem.

Finally, on the basis of the knowledge that gFCM with $\lambda \to +\infty$ reduces to bFCMA, another FCCM, referred to as bFCCMP, optimization problem is derived from the qFCCMP optimization problem with $\lambda \to +\infty$, as

$$
\begin{aligned}
\operatorname*{minimize}_{u,v,\pi} & \sum_{k=1}^{N}\sum_{i=1}^{C}(\pi_i)^{1-m}(u_{i,k})^m\left(-\ell\Gamma\left(\sum_{\ell=1}^{M}\alpha_i^{(\ell)}\right)\right. \\
& + \ell\Gamma\left(\sum_{\ell=1}^{M}\alpha_i^{(\ell)} + x_k^{(\ell)}\right) - \sum_{\ell=1}^{M}\left(\ell\Gamma\left(x_k^{(\ell)} + \alpha_i^{(\ell)}\right) + \ell\Gamma\left(\alpha_i^{(\ell)}\right)\right)\Bigg),
\end{aligned}
\tag{23}
$$

which is subject to Eqs. (6) and (7). In the ensuring subsection, we derive three FCCM algorithms based on the above optimization problems.

3.2   Proposed Algorithms

The proposed three FCCM algorithms are obtained by solving the optimization problems given in Eqs. (19), (22), and (23) subject to the constraints in Eqs. (6) and (7), where their Lagrangians $L_{\mathsf{KLFCCMP}}(u, \alpha, \pi)$, $L_{\mathsf{qFCCMP}}(u, \alpha, \pi)$,

$L_{\text{bLFCCMP}}(u, \alpha, \pi)$ are described as

$$L_{\text{KLFCCMP}}(u, \alpha, \pi) = \lambda^{-1} \sum_{k=1}^{N} \sum_{i=1}^{C} u_{i,k} \log\left(\frac{\pi_i}{u_{i,k}}\right) + \sum_{k=1}^{N} \sum_{i=1}^{C} u_{i,k} \left(\ell\Gamma\left(\sum_{\ell=1}^{M} \alpha_i^{(\ell)}\right)\right.$$

$$\left. - \ell\Gamma\left(\sum_{\ell=1}^{M} \alpha_i^{(\ell)} + x^{(\ell)}\right) + \sum_{\ell=1}^{M} \left(\ell\Gamma\left(x^{(\ell)} + \alpha_i^{(\ell)}\right) - \ell\Gamma\left(\alpha_i^{(\ell)}\right)\right)\right)$$

$$- \sum_{k=1}^{N} \gamma_k (1 - \sum_{i=1}^{C} u_{i,k}) - \eta(1 - \sum_{i=1}^{C} \pi_i), \tag{24}$$

$$L_{\text{qFCCMP}}(u, \alpha, \pi) = \frac{\lambda^{-1}}{m-1} \left(\sum_{k=1}^{N} \sum_{i=1}^{C} (\pi_i)^{1-m} (u_{i,k})^m - 1\right) + \sum_{k=1}^{N} \sum_{i=1}^{C} (\pi_i)^{1-m} (u_{i,k})^m$$

$$\times \left(-\ell\Gamma\left(\sum_{\ell=1}^{M} \alpha_i^{(\ell)}\right) + \ell\Gamma\left(\sum_{\ell=1}^{M} \alpha_i^{(\ell)} + x^{(\ell)}\right) - \sum_{\ell=1}^{M} \left(\ell\Gamma\left(x^{(\ell)} + \alpha_i^{(\ell)}\right) + \ell\Gamma\left(\alpha_i^{(\ell)}\right)\right)\right),$$

$$- \sum_{k=1}^{N} \gamma_k (1 - \sum_{i=1}^{C} u_{i,k}) - \eta(1 - \sum_{i=1}^{C} \pi_i), \tag{25}$$

$$L_{\text{bFCCMP}}(u, \alpha, \pi) = \sum_{k=1}^{N} \sum_{i=1}^{C} (\pi_i)^{1-m} (u_{i,k})^m \left(-\ell\Gamma\left(\sum_{\ell=1}^{M} \alpha_i^{(\ell)}\right)\right.$$

$$\left. + \ell\Gamma\left(\sum_{\ell=1}^{M} \alpha_i^{(\ell)} + x^{(\ell)}\right) - \sum_{\ell=1}^{M} \left(\ell\Gamma\left(x^{(\ell)} + \alpha_i^{(\ell)}\right) + \ell\Gamma\left(\alpha_i^{(\ell)}\right)\right)\right)$$

$$- \sum_{k=1}^{N} \gamma_k (1 - \sum_{i=1}^{C} u_{i,k}) - \eta(1 - \sum_{i=1}^{C} \pi_i), \tag{26}$$

respectively, with the Lagrange multipliers (γ, η). The analysis of the necessary conditions of optimality, although the detail is omitted for brevity, is summarized by the following algorithm:

Algorithm 1

STEP 1. Set the number of clusters C. Set the fuzzification parameter λ for KLFCCMP, (m, λ) for qFCCMP, and (m, λ) for bFCCMP. Set the initial object membership as u, and the initial variables controlling cluster size as π.

STEP 2. Calculate d using

$$d_{i,k} = -\ell\Gamma\left(\sum_{\ell=1}^{M} \alpha_i^{(\ell)}\right) + \ell\Gamma\left(\sum_{\ell=1}^{M} \alpha_i^{(\ell)} + x_k^{(\ell)}\right)$$

$$- \sum_{\ell=1}^{M} \left(\ell\Gamma\left(x_k^{(\ell)} + \alpha_i^{(\ell)}\right) + \ell\Gamma\left(\alpha_i^{(\ell)}\right)\right). \tag{27}$$

STEP 3. Calculate u as

$$u_{i,k} = \pi_i \exp(-\lambda d_{i,k}) / \sum_{j=1}^{C} \pi_j \exp(-\lambda d_{j,k}) \tag{28}$$

for KLFCCMP,

$$u_{i,k} = \pi_i (1 - \lambda(1-m)d_{i,k})^{1/(1-m)} / \sum_{j=1}^{C} \pi_j (1 - \lambda(1-m)d_{j,k})^{1/(1-m)} \quad (29)$$

for qFCCMP, and

$$u_{i,k} = \pi_i (d_{i,k})^{1/(1-m)} / \sum_{j=1}^{C} \pi_i (d_{j,k})^{1/(1-m)} \quad (30)$$

for bFCCMP.

STEP 4. Calculate α using the iterative formula

$$\alpha_{i,\ell} \leftarrow \alpha_{i,\ell} \frac{\sum_{k=1}^{N} u_{i,k}(x_k^{(\ell)}/(x_k^{(\ell)} - 1 + \alpha_{i,\ell}))}{\sum_{k=1}^{N} u_{i,k}(\sum_{r=1}^{M} x_k^{(r)}/(\sum_{r=1}^{M} x_k^{(r)} - 1 + \alpha_{i,r}))} \quad (31)$$

for KLFCCMP, and

$$\alpha_{i,\ell} \leftarrow \alpha_{i,\ell} \frac{\sum_{k=1}^{N} (u_{i,k})^m (x_k^{(\ell)}/(x_k^{(\ell)} - 1 + \alpha_{i,\ell}))}{\sum_{k=1}^{N} (u_{i,k})^m (\sum_{r=1}^{M} x_k^{(r)}/(\sum_{r=1}^{M} x_k^{(r)} - 1 + \alpha_{i,r}))} \quad (32)$$

for qFCCMP and bFCCMP, with a stopping criterion.

STEP 5. Calculate π as

$$\pi_i = \sum_{k=1}^{N} u_{i,k}/N \quad (33)$$

for KLFCCMP,

$$\pi_i = \frac{\left(\sum_{k=1}^{N} (u_{i,k})^m (1 - \lambda(1-m)d_{i,k}) \right)^{1/(m-1)}}{\sum_{j=1}^{C} \left(\sum_{k=1}^{N} (u_{j,k})^m (1 - \lambda(1-m)d_{j,k}) \right)^{1/(m-1)}} \quad (34)$$

for qFCCMP, and

$$\pi_i = \left(\sum_{k=1}^{N} (u_{i,k})^m d_{i,k} \right)^{1/(m-1)} / \sum_{j=1}^{C} \left(\sum_{k=1}^{N} (u_{j,k})^m d_{j,k} \right)^{1/(m-1)} \quad (35)$$

for bFCCMP.

STEP 6. Check the limiting criterion for (u, α, π). If the criterion is not satisfied, go to Step. 2.

Using the values (α, π) obtained with this algorithm, the corresponding FCF $u_i(x)$ is described as

$$u_i(x) = \pi_i \exp(-\lambda d_i(x)) / \sum_{j=1}^{C} \pi_j \exp(-\lambda d_j(x)) \quad (36)$$

for KLFCCMP,

$$u_i(x) = \pi_i(1 - \lambda(1-m)d_i(x))^{1/(1-m)} / \sum_{j=1}^{C} \pi_j(1 - \lambda(1-m)d_j(x))^{1/(1-m)} \quad (37)$$

for qFCCMP, and

$$u_i(x) = \pi_i(d_i(x))^{1/(1-m)} / \sum_{j=1}^{C} \pi_i(d_j(x))^{1/(1-m)} \quad (38)$$

for bFCCMP, where

$$d_i(x) = -\ell\Gamma\left(\sum_{\ell=1}^{M} \alpha_i^{(\ell)}\right) + \ell\Gamma\left(\sum_{\ell=1}^{M} \alpha_i^{(\ell)} + x^{(\ell)}\right)$$
$$-\sum_{\ell=1}^{M}\left(\ell\Gamma\left(x^{(\ell)} + \alpha_i^{(\ell)}\right) + \ell\Gamma\left(\alpha_i^{(\ell)}\right)\right). \quad (39)$$

4   Numerical Examples

This section presents several numerical examples that illustrate the proposed methods based on an artificial dataset.

The first dataset was obtained from the Polya mixture of two components of Polya distributions, where each component comprised 50 points on a two-dimensional simplex with the Polya parameter values $\sum_{\ell=1}^{3} x^{(\ell)} = 100$, $\alpha_1 = (3,3,12)$ and $\alpha_2 = (96,96,30)$, as shown in Fig. 1a. The conventional method, KLFCCM, and the three proposed methods, KLFCCMP, qFCCMP, and bFCCMP, with $C = 2$ and with various fuzzification parameter values, partition this dataset well, as shown in Fig. 1b, where the points in each cluster are described with circles and triangles.

The FCFs of qFCCM with $(m, \lambda) = (1 + 10^{-15}, 10^{10})$ and KLFCCM with $\lambda = 10^{10}$ are shown in Figs. 2 and 3, respectively, which are extreme cases of both fuzzification parameter values (m, λ). In these figures, the FCF values on most points are zero or one. Thus, this case shows a crisp clustering result. We find the classification borders in these figures and the difference of the classification properties between qFCCMP and KLFCCM: the classification border of qFCCMP is nonlinear, whereas that of KLFCCM is linear. This implies that qFCCMP can capture each variability of data in each cluster whereas KLFCCM cannot, which is similar to the capability of a Gaussian mixture model with different variance parameters for each cluster to capture each variance in each cluster, whereas a restricted Gaussian mixture model with unified variance parameter for all clusters cannot. Figure 2 also shows both the cases of KLFCCMP with an extreme fuzzification parameter value λ and bFCCMP with an extreme fuzzification parameter value m, because qFCCMP with $m - 1 \to +0$ reduces to KLFCCMP and qFCCMP with $\lambda \to +\infty$ reduces to bFCCMP. Then, not only qFCCMP but also KLFCCMP and bFCCMP can capture each variability of the data in each cluster. This difference between the conventional method, KLFCCM, and the three proposed methods, qFCCMP, KLFCCMP, and bFCCMP, may influence the clustering accuracy, which will be shown in later paragraphs.

The FCF of qFCCM with $(m, \lambda) = (1 + 10^{-15}, 0.5)$ is shown in Fig. 4, which is an extreme case of the fuzzification parameter value m. The difference between Figs. 2 and 4 shows that for qFCCMP, the smaller the fuzzification parameter value λ is, the fuzzier the FCF values are. Since qFCCMP with $m - 1 \to +0$ reduces to KLFCCMP, also for KLFCCMP, the smaller the fuzzification parameter value λ is, the fuzzier the FCF values are. The FCF of qFCCM with $(m, \lambda) = (1.021, 10^{10})$, which is an extreme case of the fuzzification parameter value λ, is similar to that with $(m, \lambda) = (1 + 10^{-15}, 0.5)$, shown in Fig. 4. This implies that for qFCCMP, the larger the fuzzification parameter value m is, the fuzzier the FCF values are. Although we tested qFCCMP with various fuzzification parameter values, we could not find any difference in the properties of the fuzzification parameters (m, λ) (this will be further investigated in future work). As qFCCMP with $\lambda \to +\infty$ reduces to bFCCMP, also for bFCCMP, the larger the fuzzification parameter value m is, the fuzzier the FCF values are. Revealing

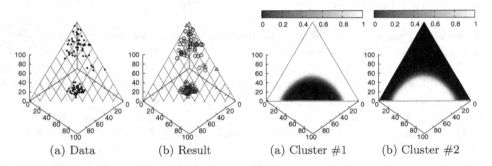

(a) Data   (b) Result   (a) Cluster #1   (b) Cluster #2

Fig. 1. Artificial dataset and its clustering result.

Fig. 2. FCFs of qFCCMP with $(m, \lambda) = (1.01, 2.0)$.

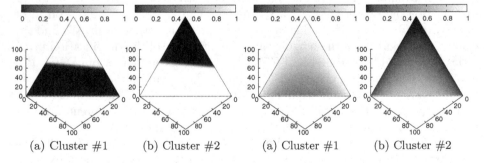

(a) Cluster #1   (b) Cluster #2   (a) Cluster #1   (b) Cluster #2

Fig. 3. FCFs of KLFCCM with $\lambda = 10$.

Fig. 4. FCFs of qFCCMP with $(m, \lambda) = (1.01, 0.5)$.

the difference among properties of the three proposed methods is also one of our future works because we could not find it. This depends on clarifying the difference in the properties of both fuzzification parameters (m, λ) in qFCCMP.

5 Conclusions

In this paper, three FCCM algorithms were proposed based on PMM and q-divergence. One algorithm was obtained by fuzzifying the log-likelihood for PMM, where as the others were obtained by modifying the first proposed algorithm based on the fact that gFCM is interpreted as a fuzzification of bFCMA by q-divergence and that gFCM connects bFCMA and eFCMA. In numerical experiments, the membership properties of the proposed methods were observed using FCF.

In future work, we will investigate the differences among the proposed three algorithms, because we could not find any in this study, in terms of both their membership properties using FCF. We will also test the proposed algorithms on many datasets and compare them with conventional methods. We also plan to apply other fuzzification techniques (e.g. [18,19]) to PMM, and extend the proposed method to possibilistic clustering [15,16] and noise clustering [17].

Acknowledgment. This work was supported by JSPS KAKENHI Grant Number 15K00348.

References

1. MacQueen, J.B.: Some methods for classification and analysis of multivariate observations. In: Proceedings of 5th Berkeley Symposium on Mathematical Statistics and Probability, pp. 281–297 (1967)
2. Bezdek, J.: Pattern Recognition with Fuzzy Objective Function Algorithms. Plenum Press, New York (1981)
3. Miyamoto, S., Mukaidono, M.: Fuzzy c-means as a regularization and maximum entropy approach. In: Proceedings of 7th International Fuzzy Systems Association World Congress (IFSA 1997), vol. 2, pp. 86–92 (1997)
4. Miyamoto, S., Kurosawa, N.: Controlling cluster volume sizes in fuzzy c-means clustering. In: Proceedings of SCIS&ISIS2004, pp. 1–4 (2004)
5. Ichihashi, H., Honda, K., Tani, N.: Gaussian mixture PDF approximation and fuzzy c-means clustering with entropy regularization. In: Proceedings of 4th Asian Fuzzy System Symposium, pp. 217–221 (2000)
6. Miyamoto, S., Ichihashi, H., Honda, K.: Algorithms for Fuzzy Clustering. Springer, Heidelberg (2008)
7. Rigouste, L., Cappé, O., Yvon, F.: Inference and evaluation of the multinomial mixture model for text clustering. Inf. Process. Manag. **43**(5), 1260–1280 (2007)
8. Dempster, A.P., Laird, N.M., Rubin, D.B.: Maximum likelihood from incomplete data via the EM algorithm. J. Royal Stat. Soc. Ser. B **39**, 1–38 (1977)
9. Honda, K., Oshio, S., Notsu, A.: FCM-type fuzzy co-clustering by K-L information regularization. In: Proceedings of FUZZ-IEEE2014, pp. 2505–2510 (2014)

10. Madsen, R.E., Kauchak, D., Elkan, C.: Modeling word burstiness using the Dirichlet distribution. In: Proceedings of ICML, pp. 545–552 (2005)
11. Chernoff, H.: A measure of asymptotic efficiency for tests of a hypothesis based on a sum of observations. Ann. Math. Stat. **23**, 493–507 (1952)
12. Tsallis, C.: Possible generalization of Boltzmann-Gibbs statistics. J. Stat. Phys. **52**, 479–487 (1988)
13. Menard, M., Courboulay, V., Dardignac, P.: Possibilistic and probabilistic fuzzy clustering: unification within the framework of the non-extensive thermostatistics. Pattern Recogn. **36**, 1325–1342 (2003)
14. Miyamoto, S., Umayahara, K.: Methods in hard and fuzzy clustering. In: Liu, Z.-Q., Miyamoto, S. (eds.) Soft Computing and Human-Centered Machines. Springer, Tokyo (2000)
15. Krishnapuram, R., Keller, J.M.: A possibilistic approach to clustering. IEEE Trans. Fuzzy Syst. **1**, 98–110 (1993)
16. Kanzawa, Y.: On possibilistic clustering methods based on Shannon/Tsallis-entropy for spherical data and categorical multivariate data. In: Torra, V., Narukawa, Y. (eds.) MDAI 2015. LNCS, vol. 9321, pp. 115–128. Springer, Cham (2015). doi:10.1007/978-3-319-23240-9_10
17. Kanzawa, Y.: On possibilistic clustering algorithms based on noise clustering. In: Proceedings of SCIS&ISIS2016, pp. 42–47 (2016)
18. Kanzawa, Y.: Generalization of quadratic regularized and standard fuzzy c-means clustering with respect to regularization of hard c-means. In: Torra, V., Narukawa, Y., Navarro-Arribas, G., Megías, D. (eds.) MDAI 2013. LNCS, vol. 8234, pp. 152–165. Springer, Heidelberg (2013). doi:10.1007/978-3-642-41550-0_14
19. Kanzawa, Y.: Power-regularized fuzzy c-means clustering with a fuzzification parameter less than one. JACIII **20**(4), 561–570 (2016)

Comparison of Fuzzy Co-clustering Methods in Collaborative Filtering-Based Recommender System

Tadafumi Kondo[✉] and Yuchi Kanzawa

Shibaura Institute of Technology, Koto, Tokyo 135-8548, Japan
ma17049@shibaura-it.ac.jp, kanzawa@sic.shibaura-it.ac.jp

Abstract. Various fuzzy co-clustering methods have been proposed for collaborative filtering; however, it is not clear which method is best in terms of accuracy. This paper proposes a recommender system that utilizes fuzzy co-clustering-based collaborative filtering and also evaluates four fuzzy co-clustering methods. The proposed system recommends optimal items to users using large-scale rating datasets. The results of numerical experiments conducted using one artificial dataset and two real datasets indicate that, the proposed method combined with a particular fuzzy co-clustering method is more accurate than conventional methods.

Keywords: Collaborative filtering · Fuzzy clustering · Co-clustering

1 Introduction

Recommender systems assist and augment our natural social process of relying on recommendations from other people to make choices without sufficient personal experience. In a typical recommender system, people provide recommendations as inputs, which the system then aggregates and directs to appropriate recipients. In some cases, the primary transformation is in the aggregation; in others, the system's value lies in its ability to make good matches between the recommenders and those seeking recommendations [1]. Recommender systems have evolved in the extremely interactive environment of the Web. They apply data analysis techniques to the problem of helping customers determine the products they would like to purchase on E-Commerce sites. For instance, a recommender system on Amazon.com (www.amazon.com) suggests items to customers based on other items customers have told Amazon they like [2].

One typical recommender system implementation method is based on collaborative filtering using evaluation data generated from users [3,4]. Collaborative filtering provides three key additional advantages to information filtering that are not provided by content-based filtering: (1) support for filtering items whose content is not easily analyzed by automated processes; (2) ability to filter items based on quality and taste; (3) ability to provide serendipitous recommendations.

© Springer International Publishing AG 2017
V. Torra et al. (Eds.): MDAI 2017, LNAI 10571, pp. 103–116, 2017.
DOI: 10.1007/978-3-319-67422-3_10

Co-clustering models have been proven useful in collaborative filtering tasks [5], where categorical multivariate datasets are provided in the form of cross-classification table, contingency table, or co-occurrence matrix. In these datasets, each individual is described by a set of qualitative variables with several categories. The categorical variables are defined by several quantifications of qualitative data: binary indicator, frequency, or scaled variable, in which several popular items can be shared by multiple clusters. Although several fuzzy co-clustering methods have been proposed [6–9], the best method for collaborative filtering in terms of accuracy is still unclear.

In this paper, we compare fuzzy co-clustering-based collaborative filtering algorithms using four fuzzy co-clustering methods specifically, FCCM [6], bFCCM [7], KLFCCM [8], and αFCCM [9] and two conventional methods specifically, Firefly [10] and GroupLens [11]. FCCM [6] is the first proposed fuzzy co-clustering model: it utilizes entropy-based fuzzification. bFCCM [7] is derived from Bezdek-type fuzzification instead of entropy-based fuzzification in FCCM. KLFCCM [8] is derived by introducing the Kullback-Leibler (KL) divergence-based regularization concept in multinomial mixture models. αFCCM [9] is derived by utilizing α-divergence instead of KL-divergence in KLFCCM. We consider two approaches to incorporate a fuzzy co-clustering method into collaborative filtering tasks: (1) using the approach proposed by Thomas and Srujana [12], in which a different type of co-clustering other than fuzzy co-clustering is used. (2) combining fuzzy co-clustering and GroupLens. The combination of two fuzzy co-clustering method collaborative filtering incorporation approaches and four fuzzy co-clustering methods yields eight fuzzy co-clustering-based collaborative filtering. Through numerical experiments using two real datasets, the proposed method with a particular fuzzy co-clustering method is shown to be more accurate than the conventional methods.

The remainder of this paper is organized as follows: Sect. 2 demonstrates user-based filtering prediction by two conventional methods and four FCCM methods. Section 3 outlines two proposed collaborative filtering algorithms that utilize co-clustering. Section 4 presents numerical experiments conducted using one artificial dataset and two real datasets. Section 5 summarizes and concludes of this paper.

2   Preliminaries

2.1   Conventional Collaborative Filtering Methods

The algorithms most frequently used in collaborative filtering are neighborhood-based methods [11]. In neighborhood-based methods, the subset of appropriate users is chosen based on their similarity to an active user and the weighted aggregate of their ratings is used to generate predictions for the active user. Let N and M be the number of users and items, respectively. Let $x \in \mathbb{R}_+^{N \times M}$ be a matrix whose (k, ℓ)-th element is the rating value of the k-th user for the ℓ-th item. Some elements of x may be missing; the goal of collaborative filtering is to predict such missing values. Define a binary matrix $y \in \mathbb{R}^{N \times M}$ by setting

$y_{k,\ell}$ equal to one if the k-th user has rated the ℓ-th item, and zero otherwise. $\hat{x}_{k,\ell}$ represents the prediction for the active user k for item ℓ. $\text{sim}(k, k')$ is the similarity weight between the active user and a neighbor k' as defined by the following Pearson correlation coefficient:

$$\text{sim}(k, k') = \frac{\sum_{\ell:y_{k,\ell}y_{k',\ell}=1}(x_{k,\ell} - \breve{x}_{k,\cdot})(x_{k',\ell} - \breve{x}_{k',\cdot})}{\sqrt{\sum_{\ell:y_{k,\ell}y_{k',\ell}=1}(x_{k,\ell} - \breve{x}_{k,\cdot})^2}\sqrt{\sum_{\ell:y_{k,\ell}y_{k',\ell}=1}(x_{k',\ell} - \breve{x}_{k',\cdot})^2}}, \quad (1)$$

where $\breve{x}_{k,\cdot}$ is the average of $\{x_{k,\ell} \mid y_{k,\ell}y_{k',\ell} = 1, \ell \in \{1, \dots, M\}\}$.

The Firefly method [10] predicts the missing values of active users, $\hat{x}_{k,\ell}$, from the other users' evaluated values and their similarities using

$$\hat{x}_{k,\ell} = \frac{\sum_{k':\text{sim}(k,k')\geq 0} x_{k',\ell}\text{sim}(k, k')}{\sum_{k':\text{sim}(k,k')\geq 0} \text{sim}(k, k')}. \quad (2)$$

The GroupLens method [11] uses Pearson correlations to weight the user similarity used by all available correlated neighbors and estimates the rating by computing the weighted average of deviations from the neighbors' mean. Then, the missing values of active users, $\hat{x}_{k,\ell}$, are predicted from the other users' evaluated values and their similarities using

$$\hat{x}_{k,\ell} = \breve{x}_{k,\cdot} + \frac{\sum_{k':\text{sim}(k,k')\geq 0} \text{sim}(k, k')(x_{k',\ell} - \breve{x}_{k',\cdot})}{\sum_{k':\text{sim}(k,k')\geq 0} \text{sim}(k, k')}, \quad (3)$$

where $x_{k',\ell}$ is replaced by $\breve{x}_{k',\cdot}$ if $x_{k',\ell}$ is missing.

Prediction methods are summarized by the following algorithm:

Algorithm 1

STEP 1 Calculate similarities using Eq. (1).
STEP 2 Calculate $\hat{x}$ using Eq. (2) for the Firefly method, and using Eq. (3) for the GroupLens method.

2.2 Fuzzy Co-clustering

Assume that for datasets $A = \{a_k \mid k \in \{1, \cdots, N\}\}$ and $B = \{b_\ell \mid \ell \in \{1, \cdots, M\}\}$, the co-occurrence information between a_k and b_ℓ, $x_{k,\ell}$ is given, where x is the matrix whose (k, ℓ)-th element is $x_{k,\ell}$. We refer to A and B as the row and column data sets, respectively, because the k-th row of x represents the similarities between a_k and b_ℓ, and the ℓ-th column of x represents the similarities between b_ℓ and a_k. The object membership of datum a_k belonging to the i-th cluster is denoted by $u_{i,k}$. The (i, k)-th element of matrix u is denoted by $u_{i,k}$, and u satisfies the constraint

$$\sum_{i=1}^{C} u_{i,k} = 1 \text{ and } u_{i,k} \in [0, 1]. \quad (4)$$

The item membership of datum b_ℓ belonging to the i-th cluster is denoted by $w_{i,\ell}$. The (i, ℓ)-th element of matrix w is denoted by $w_{i,\ell}$, and w satisfies the constraint

$$\sum_{\ell=1}^{M} w_{i,\ell} = 1 \text{ and } w_{i,\ell} \in [0, 1]. \tag{5}$$

A variable to control the i-th cluster size is denoted by π_i. The i-th element of vector π is denoted by π_i, and π satisfies the constraint

$$\sum_{i=1}^{C} \pi_i = 1. \tag{6}$$

FCCM [6] is obtained by solving the following optimization problem:

$$\operatorname*{maximize}_{u,w,\pi} \sum_{i=1}^{C} \sum_{k=1}^{N} \sum_{\ell=1}^{M} u_{i,k} w_{i,\ell} x_{k,\ell}$$

$$- \lambda_1 \sum_{i=1}^{C} \sum_{k=1}^{N} u_{i,k} \log \left(\frac{\pi_i}{u_{i,k}} \right) - \lambda_2 \sum_{i=1}^{C} \sum_{\ell=1}^{M} w_{i,\ell} \log (w_{i,\ell}), \tag{7}$$

subject to Eqs. (4), (5), and (6), where $\lambda_1 > 0$ and $\lambda_2 > 0$ are fuzzification parameters. The Bezdek-type Fuzzified FCCM method [7] is obtained by solving the following optimization problem:

$$\operatorname*{maximize}_{u,w,\pi} \sum_{i=1}^{C} \sum_{k=1}^{N} \sum_{\ell=1}^{M} (\pi_i)^{\frac{m_1-1}{m_1}} (u_{i,k})^{\frac{1}{m_1}} (w_{i,\ell})^{\frac{1}{m_2}} x_{k,\ell}, \tag{8}$$

subject to Eqs. (4), (5), and (6), where $m_1 > 1$ and $m_2 > 0$ are fuzzification parameters. KLFCCM [8] is obtained by solving the following optimization problem:

$$\operatorname*{maximize}_{u,v,\pi} \sum_{i=1}^{C} \sum_{k=1}^{N} \sum_{\ell=1}^{M} u_{i,k} \log (w_{i,\ell}) x_{k,\ell} + \lambda^{-1} \sum_{i=1}^{C} \sum_{k=1}^{N} u_{i,k} \log \left(\frac{\pi_i}{u_{i,k}} \right), \tag{9}$$

subject to Eqs. (4), (5), and (6), where $\lambda > 0$ is a fuzzification parameter. The α-divergence FCCM method [9] is obtained by solving the following optimization problem:

$$\operatorname*{maximize}_{u,v,\pi} \sum_{i=1}^{C} \sum_{k=1}^{N} \sum_{\ell=1}^{M} (\pi_i)^{1-m} (u_{i,k})^m \log(w_{i,\ell}) x_{k,\ell}$$

$$+ \frac{\lambda^{-1}}{1-m} \sum_{i=1}^{C} \sum_{k=1}^{N} (\pi_i)^{1-m} (u_{i,k})^m, \tag{10}$$

subject to Eqs. (4), (5), and (6), where $m > 1$ and $\lambda > 0$ are fuzzification parameters. The analysis of the necessary conditions of optimality (although the details are omitted for brevity) is summarized by the following algorithm:

Algorithm 2

STEP 1 Set the number of clusters as C, the fuzzification parameters (m_1, m_2) for bFCCM, (λ_1, λ_2) for FCCM, (m, λ) for KLFCCM and αFCCM, the initial cluster centers as $w_{i,\ell}$, and the initial variables controlling cluster size as π.

STEP 2 Calculate s as

$$s_{i,k} = \sum_{\ell=1}^{M} w_{i,\ell} x_{k,\ell} \tag{11}$$

for FCCM,

$$s_{i,k} = \sum_{\ell=1}^{M} (w_{i,\ell})^{\frac{1}{m_2}} x_{k,\ell} \tag{12}$$

for bFCCM and

$$s_{i,k} = \sum_{\ell=1}^{M} \log (w_{i,\ell}) x_{k,\ell} \tag{13}$$

for KLFCCM and αFCCM.

STEP 3 Calculate u as

$$u_{i,k} = \frac{\pi_i \exp \left(\lambda_1^{-1} s_{i,k}\right)}{\sum_{j=1}^{C} \pi_j \exp \left(\lambda_1^{-1} s_{j,k}\right)} \tag{14}$$

for FCCM,

$$u_{i,k} = \frac{\pi_i (s_{i,k})^{\frac{m_1}{m_1-1}}}{\sum_{j=1}^{C} \pi_j (s_{j,k})^{\frac{m_1}{m_1-1}}} \tag{15}$$

for bFCCM and

$$u_{i,k} = \frac{\pi_i \prod_{\ell=1}^{M} (w_{i,\ell})^{x_{k,\ell}\lambda}}{\sum_{j=1}^{C} \pi_j \prod_{\ell=1}^{M} (w_{j,\ell})^{x_{k,\ell}\lambda}} \tag{16}$$

for KLFCCM,

$$u_{i,k} = \frac{\pi_i (1 + \lambda(1 - m) s_{i,k})^{\frac{1}{1-m}}}{\sum_{j=1}^{C} \pi_j (1 + \lambda(1 - m) s_{j,k})^{\frac{1}{1-m}}} \tag{17}$$

for αFCCM.

STEP 4 Calculate w as

$$w_{i,\ell} = \frac{\exp \left(\lambda_2^{-1} \sum_{k=1}^{N} u_{i,k} x_{k,\ell}\right)}{\sum_{r=1}^{M} \exp \left(\lambda_2^{-1} \sum_{k=1}^{N} u_{i,k} x_{k,r}\right)} \tag{18}$$

for FCCM,

$$w_{i,\ell} = \frac{\left(\sum_{k=1}^{N}(u_{i,k})^{\frac{1}{m_1}} x_{k,\ell}\right)^{\frac{m_2}{m_2-1}}}{\sum_{r=1}^{M}\left(\sum_{k=1}^{N}(u_{i,k})^{\frac{1}{m_1}} x_{k,r}\right)^{\frac{m_2}{m_2-1}}} \tag{19}$$

for bFCCM,

$$w_{i,\ell} = \frac{\sum_{k=1}^{N} u_{i,k} x_{k,\ell}}{\sum_{r=1}^{M}\sum_{k=1}^{N} u_{i,k} x_{k,r}} \tag{20}$$

for KLFCCM and

$$w_{i,\ell} = \frac{\sum_{k=1}^{N}(u_{i,k})^{m} x_{k,\ell}}{\sum_{r=1}^{M}\sum_{k=1}^{N}(u_{i,k})^{m} x_{k,r}} \tag{21}$$

for αFCCM.

STEP 5 Calculate π as

$$\pi_i = \frac{1}{N}\sum_{k=1}^{N} u_{i,k} \tag{22}$$

for FCCM and KLFCCM,

$$\pi_i = \frac{\left(\sum_{k=1}^{N}(u_{i,k})^{\frac{1}{m_1}} s_{i,k}\right)^{\frac{1}{m_1}}}{\sum_{j=1}^{C}\left(\sum_{k=1}^{N}(u_{j,k})^{\frac{1}{m_1}} s_{j,k}\right)^{\frac{1}{m_1}}} \tag{23}$$

for bFCCM and

$$\pi_i = \frac{\left(\sum_{k=1}^{N}(u_{i,k})^{m}(1+\lambda(1-m)s_{i,\ell})\right)^{\frac{1}{m}}}{\sum_{j=1}^{C}\left(\sum_{k=1}^{N}(u_{j,k})^{m}(1+\lambda(1-m)s_{j,k})\right)^{\frac{1}{m}}} \tag{24}$$

for αFCCM.

STEP 6 Check the limiting criterion for (u, w, π). If the criterion is not satisfied, go to STEP 2.

3 Proposed Methods

We now formulate the recommendation problem in terms of a weighted matrix approximation and motivate the co-clustering approach for solving it. We applied low parameter approximations based on co-clustering of users and items in the ratings matrix x. The simplest approximation scheme based on co-clustering is one in which each missing rating is approximated by the average value in the corresponding co-cluster. We applied a more complex approximation from Thomas and Srujana [12] that incorporates the biases of the individual users and items by including the terms (user average − user-cluster average) and (item average − item-cluster average) in addition to the co-cluster average. The approximate matrix $\hat{x}_{k,\ell}$ is given by

$$\hat{x}_{k,\ell} = \bar{x}_{k,\cdot} + \bar{x}_{\cdot,\ell} - \frac{\sum_k \mu_{i,k} \bar{x}_{k,\cdot}}{\sum_k \mu_{i,k}} - \frac{\sum_\ell \omega_{i,\ell} \bar{x}_{\cdot,\ell}}{\sum_\ell \omega_{i,\ell}} + \frac{\sum_k \sum_\ell \mu_{i,k} \omega_{i,\ell} x_{k,\ell}}{\sum_k \sum_\ell \mu_{i,k} \omega_{i,\ell}}, \quad (25)$$

where $\bar{x}_{k,\cdot}$ is the average of $\{x_{k,\ell} \mid y_{k,\ell} = 1, \ell \in \{1, \ldots, M\}\}$, $\bar{x}_{\cdot,\ell}$ is the average of $\{x_{k,\ell} \mid y_{k,\ell} = 1, k \in \{1, \ldots, N\}\}$, μ is the defuzzifying of u, and ω is the defuzzifying of w. This prediction is referred to as the first proposed method. Table 1 shows an example of the evaluation values matrix before clustering, and Table 2 is a result that co-clustered the matrix, where we see four co-clusters: the first co-cluster including User #1 and #3, Item #1, #3 and #5, the second co-cluster including User #2 and #4, Item #2 and #4, the third co-cluster including User #1 and #3, Item #1, #3 and #5, the fourth co-cluster including User #2 and #4, Item #2 and #4.

Table 1. Sample of the evaluated matrix

User \ Item	1	2	3	4	5
1	$x_{1,1}$	$x_{1,2}$	$x_{1,3}$	$x_{1,4}$	$x_{1,5}$
2	$x_{2,1}$	$x_{2,2}$	$x_{2,3}$	$x_{2,4}$	$x_{2,5}$
3	$x_{3,1}$	$x_{3,2}$	$x_{3,3}$	$x_{3,4}$	$x_{3,5}$
4	$x_{4,1}$	$x_{4,2}$	$x_{4,3}$	$x_{4,4}$	$x_{4,5}$

Table 2. Sample 1 of the evaluated matrix after clustering

Cluster	User \ Item	1			2	
		1	3	5	2	4
1	1	$x_{1,1}$	$x_{1,3}$	$x_{1,5}$	$x_{1,2}$	$x_{1,4}$
	3	$x_{3,1}$	$x_{3,3}$	$x_{3,5}$	$x_{3,2}$	$x_{3,4}$
2	2	$x_{2,1}$	$x_{2,3}$	$x_{2,5}$	$x_{2,2}$	$x_{2,4}$
	4	$x_{4,1}$	$x_{4,3}$	$x_{4,5}$	$x_{4,2}$	$x_{4,4}$

During the experiments, we found that the first proposed method is inferior to the GroupLens method when applied to many datasets. Therefore, we applied the GroupLens method after fuzzy co-clustering by using the similarity with users of the user-cluster to which the active user belongs, which is referred to as the second proposed method. Let $f : X \to \{1, \ldots, C\}$ be the function indicating the index of the cluster to which the given datum belongs. Then, the missing values of active users, $\hat{x}_{k,\ell}$, from the other user's evaluated values and their similarities are obtained from

$$\hat{x}_{k,\ell} = \bar{x}_{k,\cdot} + \frac{\sum_{f(k')\equiv f(x_k)} \text{sim}(k,k')(x_{k',\ell} - \bar{x}_{k',\cdot})}{\sum_{f(k')\equiv f(x_k)} \text{sim}(k,k')}. \tag{26}$$

For example, if $x_{2,1}$ is a missing value in Table 3, Pearson's correlation coefficient between user 2 and user 4 is calculated, and the GroupLens method is applied. Assuming that Algorithm 3 is implemented, the missing values are obtained using the proposed methods.

Algorithm 3

STEP 1 Replace each missing value with the value one, which is the lowest value of ratings value.
STEP 2 Process Algorithm 2 for FCCM type clustering.
STEP 3 Calculate $\hat{x}$ as Eq. (25) for the first proposed method and Eq. (26) using for the second proposed method.

Table 3. Sample 2 of the evaluated matrix after clustering

Cluster	User	Item 1	3	5	2	4
1	1	$x_{1,1}$	$x_{1,3}$	$x_{1,5}$	$x_{1,2}$	$x_{1,4}$
	3	$x_{3,1}$	$x_{3,3}$	$x_{3,5}$	$x_{3,2}$	$x_{3,4}$
2	2	$x_{2,1}$	$x_{2,3}$	$x_{2,5}$	$x_{2,2}$	$x_{2,4}$
	4	$x_{4,1}$	$x_{4,3}$	$x_{4,5}$	$x_{4,2}$	$x_{4,4}$

4   Numerical Experiments

This section describes five example datasets used to evaluate the proposed algorithms: one artificial dataset and two real datasets.

An artificial 100×100 rating matrix composed of 100 objects and 100 items is shown in Table 4, which includes exactly 5×5 co-clusters. In the dataset, users and items #1–#20 have exactly the same ratings for all values, for the same users and items #21–#40, #61–#80, and #81–#100 have exactly the same ratings. The ideal object memberships of five object clusters is depicted in Fig. 1a, and the ideal item memberships of five item clusters is depicted in Fig. 1b, in which each row shows the 100-dimensional object membership vector $u_i = (u_{i,1}, \cdots, u_{i,100})^{\top}$ or the 100-dimensional item membership vector $w_i = (w_{i,1}, \cdots, w_{i,100})^{\top}$ by grayscale (white and black are for u_{max} (w_{max}) and 0, respectively). Then, the goal is to extract a similar structure from the dataset.

The experiment was executed as follows. The cluster numbers were set as $C \in \{4, 5, 6\}$. The fuzzification parameters were set as $(\lambda_1, \lambda_2) = (0.9, 0.8)$ for

Table 4. The artificial dataset

User \ Item	1 ⋯ 20	21 ⋯ 40	41 ⋯ 60	61 ⋯ 80	81 ⋯ 100
1	1 ⋯ 1	2 ⋯ 2	3 ⋯ 3	4 ⋯ 4	5 ⋯ 5
⋮					
20	1 ⋯ 1	2 ⋯ 2	3 ⋯ 3	4 ⋯ 4	5 ⋯ 5
21	5 ⋯ 5	1 ⋯ 1	2 ⋯ 2	3 ⋯ 3	4 ⋯ 4
⋮					
40	5 ⋯ 5	1 ⋯ 1	2 ⋯ 2	3 ⋯ 3	4 ⋯ 4
41	4 ⋯ 4	5 ⋯ 5	1 ⋯ 1	2 ⋯ 2	3 ⋯ 3
⋮					
60	4 ⋯ 4	5 ⋯ 5	1 ⋯ 1	2 ⋯ 2	3 ⋯ 3
61	3 ⋯ 3	4 ⋯ 4	5 ⋯ 5	1 ⋯ 1	2 ⋯ 2
⋮					
80	3 ⋯ 3	4 ⋯ 4	5 ⋯ 5	1 ⋯ 1	2 ⋯ 2
81	2 ⋯ 2	3 ⋯ 3	4 ⋯ 4	5 ⋯ 5	1 ⋯ 1
⋮					
100	2 ⋯ 2	3 ⋯ 3	4 ⋯ 4	5 ⋯ 5	1 ⋯ 1

(a) Ideal object membership vector u_i (b) Ideal item membership vectors w_i

Fig. 1. Ideal membership of artificial rating matrix

FCCM, $(m_1, m_2) = (1.3, 1.2)$ for bFCCM, $(\lambda = 0.1)$ for KLFCCM, and $(m, \lambda) = (1.0001, 1.0)$ for αFCCM. The initial object membership was set following the actual information. We applied the first proposed method to this dataset with the number of missing values set at $\{5, 10, 15, \ldots, 1000\}$. The rating values were chosen randomly and were caused to be missing in this dataset. For each missing value, there is a probabilistic aspect to the outcome that depends on which entries were randomly deleted. In these cases, five trials of the experiment were performed (using five different sets of incomplete data) in order to produce more significant, reproducible results. Algorithms 1 and 2 were applied to this setting. We used mean absolute error (MAE) to evaluate the prediction accuracy. MAE measures the average error in the predicted rating and the true rating. Let $x^*_{k,\ell}$ be the true ratings, and $\hat{x}_{k,\ell}$ be the ratings predicted by a recommender system. Let W be the number of user-item pairs for which the recommender system made predictions. Then, MAE is defined as follows:

$$\text{MAE} = \frac{\sum_{k=1}^{N} \sum_{\ell=1}^{M} |\hat{x}_{k,\ell} - x^*_{k,\ell}|}{W}. \tag{27}$$

The average of five MAE values for each set of missing values is shown in Fig. 2a for FCCM, Fig. 2b for bFCCM, Fig. 2c for KLFCCM, and Fig. 2d for αFCCM, where "conv.1" corresponds to Firefly, "conv.2" corresponds to GroupLens, "4C" corresponds to four clusters setting, "5C" corresponds to five clusters setting,

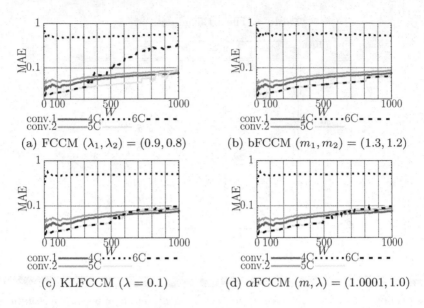

(a) FCCM $(\lambda_1, \lambda_2) = (0.9, 0.8)$

(b) bFCCM $(m_1, m_2) = (1.3, 1.2)$

(c) KLFCCM $(\lambda = 0.1)$

(d) αFCCM $(m, \lambda) = (1.0001, 1.0)$

Fig. 2. MAE on the artificial dataset using the first proposed method

and "6C" corresponds to six clusters setting. Note that every fuzzy co-clustering-based methods with both $C = 5$ and $C = 6$ exactly produce the same results except for the FCCM-based method. These figures indicate that all fuzzy co-clustering-based methods with $C = 4$ produce very bad results, all fuzzy co-clustering-based methods with $C \in \{5, 6\}$ outperform both Firefly and GroupLens for fewer missing values. In particular, the bFCCM-based method outperforms both Firefly and GroupLens for every set of missing values. From these results, it is clear that fuzzy co-clustering-based methods have the potential to outperform conventional methods if the number of clusters and initial object membership are adequately set.

The real datasets were MovieLens and BookCrossing. The MovieLens dataset was released by the GroupLens Research Project at the University of Minnesota. The data were collected through the MovieLens web site [16]. The dataset contains the response of users who were asked to rate the movies they had seen. It contains one million ratings for 3900 movies by 6040 users. Only 271,379 ratings from 905 users for 684 movies were used in this experiment; thus, each movie was evaluated by more than 240 users and each user evaluated more than 200 movies. Ratings are on a one to five scale where five is the best. The BookCrossing dataset was collected by Cai-Nicolas Ziegler in a four-week crawl (August/September 2004) of the Book-Crossing community with kind permission from Ron Hornbaker, CTO of Humankind Systems. It contains 1,149,780 ratings about 271,379 books provided by 278,858 users (anonymous but with demographic information) [17]. Only 35,157 ratings from 1090 users for 2247 books were used in this experiment. Thus, each book was evaluated by more than eight users and each user evaluated more than 15 books. Ratings were on a one to 10 scale where 10 is the best. We applied two conventional methods and two proposed methods to these two real datasets, and compared the prediction accuracy MAE, F-measure [13], and the area underneath the receiver operating characteristic (ROC) curve (AUC) [14,15].

In the proposed methods, the initial item membership values were provided in an manner similar to k-means++ [20]. Specifically, the first item membership is the normalizing value chosen uniformly at random from the data points being clustered, after which each subsequent item membership is the normalizing value chosen from the remaining data points with probability inversely proportional to its pseudo similarity from the point's closest existing initial object membership. The similarity between the i-th item membership $(w_{i,1}, \ldots, w_{i,M})$ and the k-th object x_k is calculated using Eq. (11) for FCCM, Eq. (12) for bFCCM, and Eq. (13) for KLFCCM and αFCCM. For ten initial settings, the clustering result with the maximal objective function value was selected for STEP 1 in Algorithm 3. The number of clusters and the fuzzification parameter values for each clustering algorithm were heuristically determined. They are summarized into Table 5.

The experiment was executed as follows. First, 20480 rating values were chosen randomly and caused to be missing from the all datasets, except for BookCrossing, from which only 10240 values were missing. Next, Algorithms 1 and 2 were applied to these datasets for five settings of missing values. Finally, the average of five MAE values, the average of five F-measure values, and the average of five AUC values were calculated for each dataset.

The obtained results are summarized in Tables 6, 7 and 8. Table 6 shows the MAE values for the two proposed methods and two conventional methods, Table 7 shows F-measure values for the two proposed methods and two conventional methods, and Table 8 shows AUC values for the two proposed methods and two conventional methods. The results for the first proposed method are as follows. The first proposed method is superior to both Firefly and GroupLens only for BookCrossing (Tables 6, 7, and 8). The results for the second proposed method are as follows. The second proposed method with all four clustering algorithms is superior to both Firefly and GroupLens with the case applied to BookCrossing. The second proposed method with FCCM, KLFCCM, and αFCCM is superior to both Firefly and GroupLens in terms of AUC. In particular, the second proposed method with αFCCM is superior to both Firefly and GroupLens for all three accuracy measures.

Table 5. Parameters and number of clusters set for each dataset

method / data	bFCCM (m_1, m_2)	FCCM (λ_1, λ_2)	KLFCCM (λ)	αFCCM (m, λ)	C
MovieLens	(1.001, 1.1)	(0.008, 2.0)	0.036	(1.0001, 1.6)	3
BookCrossing	(1.1, 1.1)	(0.02, 0.02)	0.0117	(1.0004, 6.4)	3

Table 6. MAE obtained for the two datasets

method / data	FireFly	GroupLens	the first proposed method				the second proposed method			
			FCCM	bFCCM	KLFCCM	αFCCM	FCCM	bFCCM	KLFCCM	αFCCM
MovieLens	0.746774	0.680339	0.692778	0.696141	0.691973	0.690771	0.680375	0.683547	0.676650	0.676493
BookCrossing	4.12383	1.30913	1.22908	1.18558	1.21262	1.29038	1.27818	1.27267	1.29851	1.26086

Table 7. F-measure obtained for the two datasets

method / data	FireFly	GroupLens	the first proposed method				the second proposed method			
			FCCM	bFCCM	KLFCCM	αFCCM	FCCM	bFCCM	KLFCCM	αFCCM
MovieLens	0.739991	0.775658	0.768426	0.766197	0.768032	0.769229	0.776292	0.774293	0.778267	0.777647
BookCrossing	0.618108	0.862537	0.871690	0.875888	0.873857	0.862651	0.867647	0.868782	0.864173	0.870518

Table 8. AUC obtained for the two datasets

method / data	FireFly	GroupLens	the first proposed method				the second proposed method			
			FCCM	bFCCM	KLFCCM	αFCCM	FCCM	bFCCM	KLFCCM	αFCCM
MovieLens	0.731914	0.791871	0.782622	0.779510	0.783315	0.784035	0.792156	0.789389	0.794649	0.794928
BookCrossing	0.454786	0.713358	0.742391	0.747041	0.746383	0.734528	0.715832	0.718077	0.715043	0.720873

5 Conclusion

In this study, we proposed two methods for a recommendation system based on fuzzy co-clustering. The results of experiments conducted on one artificial dataset and two real datasets indicate that a combination of one of the proposed methods with αFCCM is superior to conventional methods. Firefly and GroupLens in terms of accuracy metrics F-measure, AUC, and MAE.

Possible future works include adopting the deterministic annealing approach [21] by exploiting the controllable fuzzification penalty and applying various cluster validity indexes to determine the adequate number of clusters. We also plan to apply a possibilistic approach to co-clustering [22], and fuzzy co-clustering induced by multinomial mixture models (FCCMM) [23,24] in the collaborative filtering setting.

Acknowledgment. This work was supported by JSPS KAKENHI Grant Number 15K00348.

References

1. Resnick, P., Varian, H.R.: Recommender systems. In: Proceedings of Communications of the ACM, vol. 40, pp. 56–58 (1997)
2. Sarwar, B., Karypis, G., Konstan, J., Riedl, J.: Application of dimensionality reduction in recommender system - a case study. In: Proceedings of WebKDD Workshop of the ACM (2000)
3. Paul, R., Neophytos, I., Mitesh, S., Peter, S., Jhon, R.: GroupLens: an open architecture for collaborative filtering of netnews. In: Proceedings of Computer Supported Cooperative Work of the ACM, pp. 175–186 (1994)
4. Sarwar, B., Karypis, G., Riedl, J.: Item-based collaborative filtering recommendation algorithms. In: Proceedings of the 10th International Conference on World Wide Web, pp. 285–295 (2001)
5. Honda, K., Muranishi, M., Notsu, A., Ichihashi, H.: FCM-type cluster validation in fuzzy co-clustering and collaborative filtering applicability. Int. J. Comput. Sci. Netw. Secur. **13**(1), 24–29 (2013)
6. Oh, C., Honda, K., Ichihashi, H.: Fuzzy clustering for categorical multivariate data. In: Proceedings of IFSA World Congress and 20th NAFIPS International Conference, pp. 2154–2159 (2001)
7. Kanzawa, Y.: On Bezdek-type fuzzy clustering for categorical multivariate data. In: Proceedings of SCIS&ISIS, pp. 694–699 (2014)
8. Honda, K., Oshio, S., Nostu, A.: FCM-type fuzzy co-clustering by K-L information regularization. In: Proceedings of FUZZ-IEEE, pp. 2505–2510 (2014)
9. Kanzawa, Y.: Fuzzy clustering based on α-divergence for spherical data and for categorical multivariate data. In: Proceedings of FUZZ-IEEE, p. 15091 (2015)
10. Aggarwal, C.C., Wolf, J.L., Wu, W.L., Yu, P.S.: Horting hatches an egg: a new graph-theoretic approach to collaborative filtering. In: Proceedings of the 5th ACM SIGKDD International Conference on Knowledge Discovery and Data Mining, pp. 201–212 (1999)

11. Herlocker, J.L., Konstan, J.A., Borchers, A., Riedl, J.: An algorithmic framework for performing collaborative filtering. In: Proceedings of the 22nd Annual International ACM SIGIR Conference on Research and Development in Information Retrieval, pp. 230–237 (1999)
12. Thomas, G., Srujana, M.: A scalable collaborative filtering framework based on co-clustering. In: Proceedings the 5th IEEE International Conference on Data Mining, pp. 625–628 (2005)
13. Van Rijsbergen, C.J.: Information Retrieval. Butterworth-Heinemann, Newton (1979)
14. Swets, J.A.: ROC analysis applied to the evaluation of medical imaging techniques. Invest. Radiol. **14**, 109–121 (1979)
15. Hanley, J.A., McNeil, B.J.: The meaning and use of the area under a receiver operating characteristic (ROC) curve. Radiology **143**, 29–36 (1982)
16. GroupLens: MovieLens. http://grouplens.org/datasets/movielens/
17. Ziegler, C.: BookCrossing. http://www2.informatik.uni-freiburg.de/cziegler/BX/
18. Goldberg, K.: Jester. http://eigentaste.berkeley.edu/dataset/
19. Petricek, V.: LibimSeTi. http://www.occamslab.com/petricek/data/
20. Arthur, D., Vassilvitskii, S.: k-means++: the advantages of careful seeding. In: Proceedings of the 8th Annual ACM-SIAM Symposium on Discrete Algorithms, pp. 1027–1035 (2007)
21. Rose, K., Gurewitz, E., Fox, G.: A deterministic annealing approach to clustering. Pattern Recognit. Lett. **11**, 589–594 (1990)
22. Kanzawa, Y.: On possibilistic clustering methods based on Shannon/Tsallis-entropy for spherical data and categorical multivariate data. In: Torra, V., Narukawa, Y. (eds.) MDAI 2015. LNCS, vol. 9321, pp. 115–128. Springer, Cham (2015). doi:10.1007/978-3-319-23240-9_10
23. Honda, K.: Fuzzy co-clustering and application to collaborative filtering. In: Huynh, V.-N., Inuiguchi, M., Le, B., Le, B.N., Denoeux, T. (eds.) IUKM 2016. LNCS, vol. 9978, pp. 16–23. Springer, Cham (2016). doi:10.1007/978-3-319-49046-5_2
24. Nakano, T., Honda, K., Ubukata, S., Notsu, A.: A study on recommendation ability in collaborative filtering by fuzzy co-clustering with exclusive item partition. In: 2016 Joint 8th International Conference on Soft Computing and Intelligent Systems (SCIS) and 17th International Symposium on Advanced Intelligent Systems (ISIS), pp. 686–689 (2016)

Data Privacy and Security

Differentially Private Data Sets Based on Microaggregation and Record Perturbation

Jordi Soria-Comas$^{(\boxtimes)}$ and Josep Domingo-Ferrer

UNESCO Chair in Data Privacy, Department of Computer Science and Mathematics,
Universitat Rovira i Virgili, Av. Països Catalans 26,
43007 Tarragona, Catalonia, Spain
{jordi.soria,josep.domingo}@urv.cat

Abstract. We present an approach to generate differentially private data sets that consists in adding noise to a microaggregated version of the original data set. While this idea has already been proposed in the literature to reduce the data sensitivity and hence the noise required to reach differential privacy, the novelty of our approach is that we focus on the microaggregated data set as the target of protection, rather than focusing on the original data set and viewing the microaggregated data set as a mere intermediate step. As a result, we avoid the complexities inherent to the insensitive microaggregation used in previous contributions and we significantly improve the utility of the data. This claim is supported by theoretical and empirical utility comparisons between our approach and existing approaches.

Keywords: Anonymization · Differential privacy · Microaggregation · Privacy

1 Introduction

Microdata (that is, information at the individual level) are usually the most convenient type of data for secondary use. However, the risk of disclosure inherent to releasing such detailed information is significant. Traditionally, data were mostly handled by a reduced number of data controllers (e.g. national statistical offices), who had collected them under strong pledges of privacy. In that scenario, reasonable assumptions about the knowledge available to intruders could be made and the methodology for disclosure risk limitation could be adjusted accordingly. Nowadays, the developments in information technology facilitate the collection of personal data. This bounty of data makes it increasingly difficult to make well-grounded assumptions about the side knowledge available to potential intruders [1].

Differential privacy [2] (DP) is a well-known privacy model that gives privacy guarantees without making any assumption on the intruder's side knowledge. In this sense, DP suits well the current scenario with many data controllers. Unlike privacy models designed to protect sets of microdata (e.g. k-anonymity [3],

© Springer International Publishing AG 2017
V. Torra et al. (Eds.): MDAI 2017, LNAI 10571, pp. 119–131, 2017.
DOI: 10.1007/978-3-319-67422-3_11

l-diversity [4], t-closeness [5]), DP was designed to protect the outcomes of interactive queries. However, this limitation was soon overcome with the development of several approaches to release differentially private microdata (DP microdata) [6–10].

The dominant approach to generate DP microdata is based on the computation of DP histograms. However, histogram-based approaches have severe limitations when the number of attributes grows: for fixed attribute granularities, the number of histogram bins grows exponentially with the number of attributes, which has a severe impact on both computational cost and accuracy. To avoid these issues, we propose to generate the DP data set by masking the records in the original data set. Plain independent masking of the records in the original data set is computationally very efficient (its cost is linear on the size of the data set). However, the amount of masking needed to achieve DP is proportional to the sensitivity (the maximum possible variation) of what is being masked, and the sensitivity of an attribute value in a record is large (typically, as large as the attribute domain size). Therefore, a large amount of masking is needed, that results in very substantial information loss.

In this work we describe a record-level perturbation-based approach to generate DP data sets that uses microaggregation to reduce the sensitivity of attribute values and hence the amount of noise required to attain DP. Our approach does not require the use of any specific microaggregation algorithm, but we will choose some microaggregation algorithms for the sake of evaluation. We also compare our results to previous record perturbation approaches. In Sect. 2 we briefly introduce some basic concepts about DP. In Sect. 3 we describe our approach to generate DP data sets. In Sect. 4 we evaluate several microaggregation strategies theoretically and experimentally (by comparing results among them and by comparing results to already existing approaches). Finally, in Sect. 5 we summarize the conclusions and outline future research avenues.

2 Background on Differential Privacy

Differential privacy [2] is popular among academics due to the strong privacy guarantees it offers. DP does not rely on assumptions about the side knowledge available to the intruders. Rather, disclosure risk limitation is tackled in a relative manner: the result of any analysis should be similar between data sets that differ in one record. As stated in [11], under DP individuals have no privacy reason to refuse participating in a data set:

> Any given disclosure will be, within a multiplicative factor, just as likely whether or not the individual participates in the database. As a consequence, there is a nominally higher risk to the individual in participating, and only nominal gain to be had by concealing or misrepresenting one's data.

Differential privacy assumes the presence of a trusted party that: (i) holds the data set, (ii) receives the queries submitted by the data users, and (iii) responds to them in a privacy-aware manner. The notion of differential privacy is formalized according to the following definition:

Definition 1 (ϵ-differential privacy). *A randomized function κ gives ϵ-differential privacy (ϵ-DP) if, for all data sets D_1 and D_2 that differ in one record (a.k.a. neighbor data sets), and all $S \subset Range(\kappa)$, we have*

$$\Pr(\kappa(D_1) \in S) \leq \exp(\epsilon) \Pr(\kappa(D_2) \in S).$$

Given a query function f, the goal in differential privacy is to find a randomized function κ_f that satisfies ϵ-DP and approximates f as closely as possible. For the case of numerical queries, κ_f can be obtained via noise addition; that is $\kappa_f(\cdot) = f(\cdot) + N$, where N is a random noise that has been properly adjusted to attain ϵ-DP. The addition of a Laplace distributed noise whose scale has been adjusted to the global sensitivity of the query f is, probably, the most common approach (although other approaches has been proposed [12–14]).

Definition 2 (L_1-sensitivity). *The L_1-sensitivity, Δf, of a function $f : \mathcal{D}^n \to \mathbb{R}^d$ is the maximum variation of f between data sets that differ in one record:*

$$\Delta f = \max_{d(D,D')=1} \|f(D) - f(D')\|_1.$$

Proposition 1. *Let $f : \mathcal{D}^n \to \mathbb{R}^d$ be a function. The mechanism $\kappa_f(D) = f(D) + (N_1, \ldots, N_d)$, where N_i are drawn i.i.d. from a Laplace$(0, \Delta f/\epsilon)$ distribution, is ϵ-DP.*

3   DP Data Sets via Microaggregation

Let D be the collected data set. Assume that we want to generate D^ϵ –an anonymized version of D– that satisfies ϵ-DP. Let $I_r(D)$ be the query that returns r. We can think of the data set D as the collected answers to the queries $I_r(D)$ for $r \in D$, and we can generate D_ϵ by collecting ϵ-DP responses to $I_r(D)$ for $r \in D$. Such a naive procedure to generate a DP data set is, however, likely to produce a large information loss. In the end, the purpose of DP is to make sure that individual records do not have any significant effect on query responses, which implies that the accuracy of the responses to $I_r(D)$ is necessarily low.

To make perturbative masking viable for the generation of DP data sets, we have to reduce the sensitivity of the queries used. This requires a shift from individual queries to queries that ask for aggregate or statistical information. Along the lines of [9,10,15,16], our proposal is based on microaggregation. In spite of microaggregation being itself a well-known technique in disclosure risk limitation, we use it here with the sole purpose of reducing the sensitivity of the queries. The disclosure risk limitation comes from the enforcement of DP.

This change of purpose carries along a change in the traditional way of thinking about microaggregation.

In standard microaggregation, one splits the data set into clusters of at least k records and then replaces the records in each cluster by the cluster centroid, where the minimum value k prevents the cluster from being too representative of any individual in it. In our case, we are also interested in having not too small clusters (in order to limit the impact of individual contributions and hence the sensitivity), but we can relax the requirement of a minimum cluster size. In our case, the total error is the combination of the error introduced by microaggregation and the error due to noise addition; thus, if adding one more record to a cluster produces a large increase in the microaggregation error, it may be preferable to use the smaller cluster. In this work, we think of microaggregation as an algorithm that proceeds in the following two steps:

1. Split the data set into clusters of records.
2. Compute a representative record of each cluster and replace the records in the cluster by it.

To reduce the error introduced by microaggregation, we usually want to generate clusters that are as homogeneous as possible. For the sake of generality, in this section, we do not favor any particular strategy to generate the microaggregation clusters: they can all have the same cardinality or different ones, they can be optimal (maximally homogeneous) or not, randomized or deterministic, etc. However, to be able to analyze the effect of microaggregation on the sensitivity, we need to fix the particular way in which the records in a cluster are combined to generate a record that is representative of the cluster. In this work, we use the mean as aggregation operation (that is, we compute the centroid of the cluster).

The approach we propose is different from those of [9,10,15,16], in that here we consider that *the data set to be protected is the microaggregated one*, rather than the original one. In other words, given an original data set D, we generate $\bar{D}$ by microaggregation of the records in D. From this point on, we discard D and we focus on protecting $\bar{D}$. Hence, the goal is to publish $\bar{D}^{\epsilon}$, a DP version of $\bar{D}$.

The data set $\bar{D}$ acts as a proxy of the original data set D. Thus, when evaluating the utility of $\bar{D}^{\epsilon}$ we need to account for two sources of error: (i) the error due to the microaggregation (that is, the error caused by using $\bar{D}$ as a proxy of D), and (ii) the noise introduced to attain ϵ-DP. The advantage of the proposed approach lies in the fact that the error introduced in the microaggregation step is likely to be more than compensated by the reduction in the noise required to attain DP (compared to the noise that would be required to attain DP directly from the original data set D).

Since the contribution of a record to the centroid is inversely proportional to the cardinality of the corresponding cluster, the centroid sensitivity can be obtained as the record sensitivity divided by the cluster cardinality. This is formalized in the following proposition.

Proposition 2. *Let $C \subset D$ be a cluster of records and let c be the mean of the records in C. Let ΔD be the L_1-sensitivity of a record in D. The L_1-sensitivity of the centroid c is $\Delta c = \Delta D/|C|$.*

Proof. Δc represents the maximum change in c due to an arbitrary change in a single record. Since the maximum change in a single record is ΔD and each record contributes to c, at most, in a proportion of $1/|C|$, the maximum change in c is $\Delta D/|C|$. ☐

Notice that the sensitivities may differ for centroids of different clusters, because the sensitivity depends on the cluster cardinality. Once the sensitivity of a centroid c is computed, ϵ-DP can be attained by adding a Laplace noise with zero mean and scale $\Delta c/\epsilon$. Since each cluster contains disjoint records, parallel composition applies; thus, by adding Laplace noise independently to each cluster, we obtain the list of ϵ-DP centroids (see Fig. 1).

Since each record replaced by the corresponding centroid, each centroid is repeated as many times as there are records in the corresponding cluster. We now explain why in Fig. 1 all repetitions of a centroid value are added exactly the same noise. If we added a different random noise to each repetition of the centroid, we would have $|C|$ non-independent DP outcomes each of which has sensitivity $\Delta D/|C|$; hence, by sequential composition, the sensitivity of the list of centroid repetitions in the cluster would be ΔD, which would cancel the benefits of microaggregation. To keep the sensitivity of the centroid repetitions at $\Delta D/|C|$, we must have a single DP centroid value, that is, we must add exactly the same noise to all the repetitions of given centroid. In other words, for each cluster C_i, we take a single draw, n_i, from the $Laplace(0, \frac{\Delta D}{|C_i|\epsilon})$ distribution and use it to mask the $|C_i|$ occurrences of c_i.

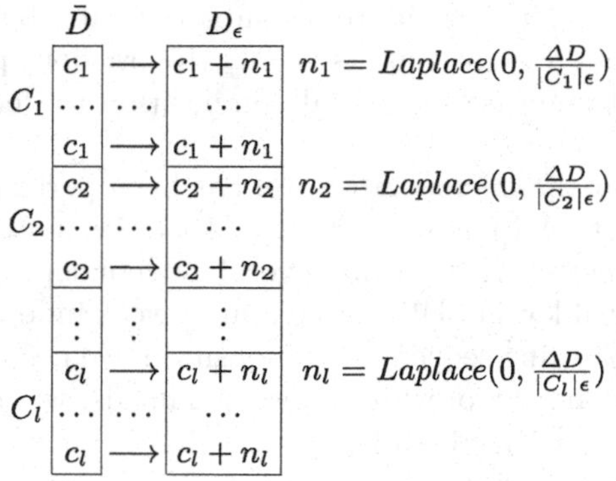

$$
\begin{array}{c|ccc|c|c}
 & \bar{D} & & & D_\epsilon & \\
\hline
 & c_1 & \longrightarrow & c_1 + n_1 & n_1 = Laplace(0, \frac{\Delta D}{|C_1|\epsilon}) \\
C_1 & \cdots & \cdots & \cdots & \\
 & c_1 & \longrightarrow & c_1 + n_1 & \\
\hline
 & c_2 & \longrightarrow & c_2 + n_2 & n_2 = Laplace(0, \frac{\Delta D}{|C_2|\epsilon}) \\
C_2 & \cdots & \cdots & \cdots & \\
 & c_2 & \longrightarrow & c_2 + n_2 & \\
\hline
 & \vdots & \vdots & \vdots & \\
\hline
 & c_l & \longrightarrow & c_l + n_l & n_l = Laplace(0, \frac{\Delta D}{|C_l|\epsilon}) \\
C_l & \cdots & \cdots & \cdots & \\
 & c_l & \longrightarrow & c_l + n_l & \\
\end{array}
$$

Fig. 1. Generation of an ϵ-DP data set using record-level microaggregation to reduce the amount of noise required

The procedure to generate an ϵ-DP data set based on record-level microaggregation is formally described in Algorithm 1. The algorithm takes as input parameters the microaggregated data set $\bar{D}$ (whose records consist of the corresponding

Algorithm 1. Procedure to generate an ϵ-DP data set using record-level microaggregation to reduce the amount of noise required

Require:

$\bar{D} = \{r_1, \ldots, r_L\}$: microaggregated data set (each record r_j is the corresponding cluster centroid)

Mapping τ between records of $\bar{D}$ and the clusters $C_1, \ldots, C_l$ formed in the microaggregation

ϵ: desired level of DP

Output

$\bar{D}^\epsilon$: an ϵ-DP data set

for $i \in \{1, \ldots, l\}$ **do**

    set $n_i =$ random draw from the $Laplace(0, \frac{\Delta D}{|C_i|\epsilon})$ distribution

end for

for $j \in \{1, \ldots, L\}$ **do**

    let $C_i := \tau(r_j)$

    set $r_j^\epsilon = r_j + n_i$

end for

return $\bar{D}^\epsilon = \{r_1^\epsilon, \ldots, r_L^\epsilon\}$

---

cluster centroids), the mapping between records in $\bar{D}$ and clusters, and the desired level ϵ of DP. Next, we fix the noise n_i that will be added to all records mapped to each cluster C_i. Finally, we loop through the records in $\bar{D}$ and add to each record the noise that corresponds to the cluster it is mapped to.

The procedure depicted in Fig. 1 assumes that microaggregation is performed over whole records (either because the data set contains a single attribute or because multivariate microaggregation over all the attributes is used). In the remainder of this section, we generalize the previous procedure to work independently with several individual attributes or subsets of attributes. Essentially, we split the attributes into disjoint subsets, apply the previous procedure independently to each subset, and use sequential composition to determine the overall level of DP.

Let us assume that the microaggregation has been performed independently over the disjoint subsets of attributes $AS_1, \ldots, AS_m$. Sequential composition says that the level of differential privacy from several independent queries accumulates to determine the overall level of DP. As we aim to work independently with each of the subsets AS_i, following sequential composition, we need to split the overall privacy budget, ϵ, among the previous subsets. That is, we fix values $\epsilon_1, \ldots, \epsilon_m$ subject to the restrictions $\epsilon_i \geq 0$ and $\epsilon_1 + \ldots + \epsilon_m = \epsilon$. For each subset AS_i, we apply the procedure in Algorithm 1 to attain ϵ_i-DP. Sequential composition tells that the result is ϵ-DP. This is illustrated in Fig. 2 and formalized in Algorithm 2.

4   Evaluation

We evaluate the proposal in Sect. 3 by fixing several microaggregation strategies and comparing the new proposal to existing methods that are also based on

$$\begin{array}{cc} \bar{D} & \bar{D}_\epsilon \end{array}$$

$$\begin{array}{ccc} AS_1 & \cdots & AS_m \end{array} \qquad \begin{array}{ccc} AS_1 & \cdots & AS_m \end{array}$$

$$\begin{vmatrix} c^1_{\rho_1(1)} & \cdots & c^m_{\rho_m(1)} \\ c^1_{\rho_1(2)} & \cdots & c^m_{\rho_m(2)} \\ \vdots & \vdots & \vdots \\ c^1_{\rho_1(n)} & \cdots & c^m_{\rho_m(n)} \end{vmatrix} \begin{array}{c} \longrightarrow \\ \longrightarrow \\ \vdots \\ \longrightarrow \end{array} \begin{vmatrix} c^1_{\rho_1(1)} + n^1_{\rho_1(1)} & \cdots & c^m_{\rho_m(1)} + n^m_{\rho_m(1)} \\ c^1_{\rho_1(2)} + n^1_{\rho_1(2)} & \cdots & c^m_{\rho_m(2)} + n^m_{\rho_m(2)} \\ \vdots & \vdots & \vdots \\ c^1_{\rho_1(n)} + n^1_{\rho_1(n)} & \cdots & c^m_{\rho_m(n)} + n^m_{\rho_m(n)} \end{vmatrix}$$

where $\rho_i(r)$ =cluster number associated to record r

$$n^i_j = Laplace(0, \tfrac{\Delta AS_i}{(|C^i_j|\epsilon_i})$$

Fig. 2. Generation of an ϵ-DP data set by independently microaggregating the subsets of attributes $AS_1, \ldots, AS_m$ and reaching ϵ_i-DP for group AS_i

---

Algorithm 2. Procedure to generate an ϵ-DP data set by independently microaggregating the groups of attributes $AS_1, \ldots, AS_n$ and reaching ϵ_i-DP for group AS_i

---

Require:

$AS_1, \ldots, AS_m$: list of disjoint subsets of attributes

$\bar{D}$: microaggregated data set, where microaggregation has been independently computed for the projections on each subset of attributes (each record has been replaced by the centroids of the clusters that contain it in each projection)

$(\tau_1, \ldots, \tau_m)$: τ_i is the mapping between records in $\bar{D}$ and the clusters $C^i_1, \ldots, C^i_{l_i}$ computed for the projection $\bar{D}[AS_i]$ of $\bar{D}$ on attribute subset AS_i

$\epsilon_1, \ldots, \epsilon_m$: level of DP for attributes AS_i (subject to $\sum \epsilon_i = \epsilon$)

Output

$\bar{D}^\epsilon$: an ϵ-DP data set

for $i \in \{1, \ldots, m\}$ **do**

$\quad \bar{D}^\epsilon[AS_i] = Algorithm\ 1(\bar{D}[AS_i], \tau_i, \epsilon_i)$

end for

return $\bar{D}^\epsilon$

---

record perturbation [9,10]. At first sight, the fact that we employ basic microaggregation algorithms rather than (the more restrictive and less utility-preserving) insensitive microaggregation [9] seems a substantial advantage. Moreover, the method in Sect. 3 allows adjusting the noise to the size of each cluster.

A difference between the method of Sect. 3 and the methods in [9,10] is that the former considers that the data set to be protected is the microaggregated one ($\bar{D}$), whereas the latter aim at protecting the original data set (D). Nonetheless, regardless of the method used, utility must be evaluated in terms of how good is the DP data set D^ϵ as a replacement for the original data set D.

4.1   Evaluated Methods

In Sect. 3 we did not favor any microaggregation strategy. However, the fact is that the microaggregation approach has a significant impact on the utility of the DP data set output by our method. For that reason, empirical results are necessarily tied to a specific microaggregation strategy.

We evaluate the accuracy of our proposal when microaggregation is instantiated with the MDAV algorithm (a heuristic multivariate microaggregation algorithm, [17]) and with individual ranking MDAV microaggregation (which runs independent univariate MDAV microaggregations for each attribute). We have chosen these microaggregation algorithms not only because they are well known, but because they have previously been used to improve the accuracy of DP data sets generated via record perturbation [9,10].

It is clear, however, that the above-mentioned microaggregation algorithms have some restrictions that limit the accuracy improvements they can offer. An important limitation is that the clusters they generate have a fixed cardinality k (except, maybe, the last cluster, that is of size between k and $2k - 1$). However, as noted in Sect. 3, the method to generate DP data sets described in that section does not require a fixed cluster size, not even a minimum cluster size.

We have evaluated the following DP methods in our comparison:

- MDAV+DP. The method described in Sect. 3 instantiated with a multivariate MDAV microaggregation of entire records.
- IR_MDAV+DP. The method described in Sect. 3 instantiated with individual ranking MDAV microaggregation.
- INS+DP (baseline). The method for DP based on insensitive multivariate microaggregation that is described in [9]. This method is a suitable comparison baseline for MDAV+DP because both methods use multivariate microaggregation of entire records.

The method described in [10] could also be considered as a comparison baseline (it would be a good baseline for IR_MDAV+DP, because both are based on individual ranking MDAV microaggregation). However, we skip it because the computation of the sensitivity in [10] is flawed, which leads to overly reducing the noise required to attain DP.

Even if they do not yield DP, the standalone MDAV and IR_MDAV microaggregation algorithms (without subsequent noise addition to attain DP) have also been evaluated. The reason is that using standalone MDAV and IR_MDAV provides an upper bound of the accuracy reachable with MDAV+DP and IR_MDAV+DP, respectively.

4.2 Theoretical Evaluation

Although an empirical evaluation is provided further below, we think that a theoretical comparison of some methods, specifically MDAV+DP and IR_MDAV+DP, can yield some important insights.

The following proposition shows that both MDAV+DP and IR_MDAV+DP can yield an ϵ-DP data set by adding the same amount of noise to each attribute.

Proposition 3. *Given a cluster size k used in microaggregation and a target DP level ϵ, both MDAV+DP and IR_MDAV+DP can yield an ϵ-DP data set by adding the same amount of noise to each original attribute.*

Proof. According to Algorithm 1, to attain ϵ-DP with MDAV+DP, we need to add a noise that is distributed according to a $Laplace(0, \Delta D/\epsilon)$ to each attribute. Assume now we use IR_MDAV+DP instead, and attribute i having sensitivity ΔD_i is added noise drawn from $Laplace(0, \Delta D_i/\epsilon_i)$. Both Laplace distributions are equal when $\Delta D/\epsilon = \Delta D_i/\epsilon_i$, which may be enforced by taking

$$\epsilon_i = \epsilon \frac{\Delta D_i}{\Delta D}. \tag{1}$$

Since $\Delta D = \sum \Delta D_i$, the sum of the ϵ_i amounts to ϵ (as required by the IR_MDAV+DP method). $\qquad\square$

The conclusion from the previous proposition is that IR_MDAV+DP (with appropriate ϵ_i) should always be preferred to MDAV+DP: the error due to microaggregation is smaller with IR_MDAV+DP (because less attributes are clustered together) and the error due to noise addition can be made equal. In spite of this result, for the sake of completeness, we will perform the empirical evaluation over both IR_MDAV+DP and MDAV+DP. Actually, we consider two variants of IR_MDAV+DP: IR_MDAV+DP_1 uses the same level of DP for all attributes ($\epsilon_1 = \ldots = \epsilon_m = \epsilon/m$), and IR_MDAV+DP_2 uses the values of ϵ_i given by Expression (1), for $i = 1, \ldots, m$, so that Proposition 3 holds.

4.3   Evaluation Data

The empirical evaluation has been performed on the Census data set, which was first used in the "CASC" European project [18] as a reference data set to test and compare statistical disclosure control methods, and was also used in [9]. This data set contains 13 numerical attributes and 1080 records. For the sake of comparability with [9], we focus on 4 attributes: FICA (Social security retirement payroll deduction), FEDTAX (Federal income tax liability), INTVAL (Amount of interest income) and POTHVAL (Total other persons income).

The selected attributes take values above 0 but they are not naturally upper-bounded. Since the L_1-sensitivity is proportional to the sizes of the domains of attributes, we need to upper-bound the domain of each attribute. For the sake of comparability, we use the upper bounds that were used in [9]; that is, we upper-bound the domain of an attribute by 1.5 times the maximum value of the attribute in the data set. The domain bounds on the attributes are also enforced when adding noise to attain DP: the DP masked values are truncated to lie within the fixed bounds.

4.4   Evaluation Measures

The evaluation is based on two measures of error: the sum of squared errors (SSE) and the sum of absolute errors (SAE). The SSE is a measure of overall information loss that is commonly used in the evaluation of SDC methods (and particularly in microaggregation). It is computed as

$$SSE = \sum_{i=1,\ldots,n} \sum_{j=1,\ldots,m} (r_{ij} - r_{ij}^{\epsilon})^2$$

where r_{ij} is the value of attribute j in original record r_i and r_{ij}^{ϵ} is value of attribute j in the record r_i^{ϵ} of the DP data set $\bar{D}^{\epsilon}$ that corresponds to r_i.

SAE is similar to SSE but, rather than being based on squared errors, it is based on absolute errors. It is computed as

$$SAE = \sum_{i=1,\ldots,n} \sum_{j=1,\ldots,m} |r_{ij} - r_{ij}^{\epsilon}|.$$

Both measures give an overall estimation of the error in the generated data set but they differ in the relative importance they attach to the magnitude of each difference. In SSE a large error in a single record may have a large overall impact, while in SAE a large error in a single record can be more easily compensated by small errors in other records.

4.5   Experimental Results

Figure 3 shows the evolution of SSE as a function of the cluster size. In both graphs of the figure we can see that, as expected, the SSE for the microaggregation algorithms MDAV and IR_MDAV increases with the size of the cluster (which is represented in the abscissae). There is a steep increase for small cluster sizes that flattens out progressively as the cluster size gets larger. On the contrary, for MDAV+DP and IR_MDAV+DP the opposite occurs: SSE decreases with the size of the clusters and the decrease is steeper for small cluster sizes. We observe that, for large cluster sizes, the SSE of all DP methods converge to the SSE of the underlying microaggregation. This result was to be expected because, the greater the cluster size, the less noise we need to attain DP. As it can be seen by comparing both graphics, the rate of convergence is proportional to ϵ (faster convergence for larger ϵ). The comparison between MDAV_DP and IR_MDAV+DP (both variants) shows that IR_MDAV+DP has a lower SSE. This could also be expected, because IR_MDAV is more utility-preserving than MDAV. The comparison between IR_MDAV+DP_1 and IR_MDAV+DP_2 shows that IR_MDAV+DP_2 has slightly less SSE than IR_MDAV+DP_1, but the difference seems to be relatively small.

We then compared the SSE obtained with the methods in this paper with the SSE obtained with the method in [9]. Figure 4a in [9] shows the SSE of the DP data set generated by performing a prior insensitive microaggregation to reduce the noise needed to reach DP. By comparing that figure with Fig. 3, we observe that IR_MDAV+DP with $\epsilon = 1$ performs as well as the insensitive approach in [9] with $\epsilon = 10$. This is a very significant improvement in the utility of the data.

Figure 4 shows the SAE of MDAV+DP, IR_MDAV+DP_1 and IR_MDAV+DP_2, and compares them with the baseline MDAV and IR microaggregation algorithms. Consistently with the theoretical comparison between MDAV+DP and

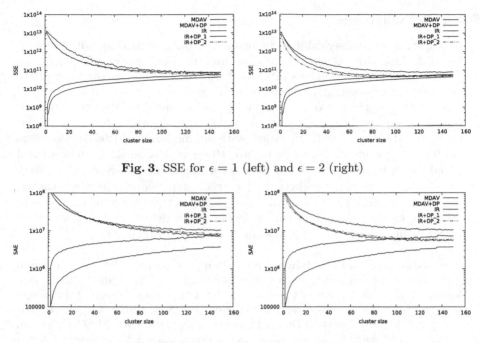

Fig. 3. SSE for $\epsilon = 1$ (left) and $\epsilon = 2$ (right)

Fig. 4. SAE for $\epsilon = 1$ (left) and $\epsilon = 2$ (right)

IR_MDAV+DP above and with the SSE results, we observe that IR_MDAV+DP is more utility-preserving.

5   Conclusions and Future Work

We have presented an approach to generate DP data sets that consists of adding noise to a microaggregated version of the original data set. Using microaggregation as a prior step to reduce the sensitivity of the data and hence the noise that needs to be added to reach DP had already been proposed in the literature. However, the novelty of our approach is that we focus on the microaggregated data set as the target of protection, rather than focusing on the original data set and viewing the microaggregated data set as a mere intermediate step. As a result, we avoid the complexities inherent to insensitive microaggregation and significantly improve the utility of the data.

The approach we have presented works with any microaggregation algorithm. For concreteness and convenience, we have analyzed three actual approaches to generate DP data sets: MDAV_DP and two variants of IR_MDAV_DP. The comparison (both theoretical and empirical) has shown that IR_MDAV_DP is better than MDAV_DP. Comparisons of IR_MDAV_DP with the insensitive based approach in [9] have shown that IR_MDAV_DP with $\epsilon = 1$ is similar in terms of SSE to the insensitive approach with $\epsilon = 10$. This is a significant improvement in the utility with respect to prior work.

Future work will include:

– Considering non-numerical data by using microaggregation algorithms capable of dealing with categorical data (ordinal, nominal or hierarchical).
– Trying aggregation operators different from the mean (e.g. the medoid) to compute the representative record of a cluster.
– Using variable-size microaggregation heuristics, such as [19], without minimum cluster size (that is, taking $k = 1$). The optimal solution to standalone variable-size microaggregation without minimum cluster size consists of all clusters containing a single record. However, the optimal solution when variable-size microaggregation is used as a preliminary step of DP is likely to contain larger clusters (because larger clusters reduce the noise that is needed to attain DP). In general, the less restrictive nature of variable-size microaggregation algorithms can be expected to deliver DP data sets with better utility, at the cost of increasing the computational effort.

Acknowledgments and Disclaimer. Partial support to this work has been received from the European Commission (projects H2020-644024 "CLARUS" and H2020-700540 "CANVAS"), the Government of Catalonia (ICREA Acadèmia Prize to J. Domingo-Ferrer), and from the Spanish Government (projects TIN2014-57364-C2-1-R "Smart-Glacis" and TIN 2015-70054-REDC). The authors are with the UNESCO Chair in Data Privacy, but the views in this paper are their own and are not necessarily shared by UNESCO.

References

1. Soria-Comas, J., Domingo-Ferrer, J.: Big data privacy: challenges to privacy principles and models. Data Sci. Eng. **1**(1), 21–28 (2015)
2. Dwork, C., McSherry, F., Nissim, K., Smith, A.D.: Calibrating noise to sensitivity in private data analysis. In: Halevi, S., Rabin, T. (eds.) TCC 2006. LNCS, vol. 3876, pp. 265–284. Springer, Heidelberg (2006). doi:10.1007/11681878_14
3. Samarati, P., Sweeney, L.: Protecting privacy when disclosing information: k-anonymity and its enforcement through generalization and suppression. Technical report, SRI International (1998)
4. Machanavajjhala, A., Kifer, D., Gehrke, J., Venkitasubramaniam, M.: l-diversity: privacy beyond k-anonymity. ACM Trans. Knowl. Disc. Data **1**(1) (2007).
5. Li, N., Li, T., Venkatasubramanian, S.: t-closeness: privacy beyond k-anonymity and l-diversity. In: 23th IEEE International Conference on Data Engineering-ICDE 2007, pp. 106–115. IEEE (2007)
6. Machanavajjhala, A., Kifer, D., Abowd, J., Gehrke, J., Vilhuber, L.: Privacy: theory meets practice on the map. In: 24th IEEE International Conference on Data Engineering-ICDE 2008, pp. 277–286 (2008)
7. Zhang, J., Cormode, G., Procopiuc, C.M., Srivastava, D., Xiao, X.: Privbayes: private data release via Bayesian networks. In: 2014 ACM SIGMOD International Conference on Management of Data-SIGMOD 2014, pp. 1423–1434. ACM, New York (2014)
8. Xiao, Y., Xiong, L., Yuan, C.: Differentially private data release through multidimensional partitioning. In: Jonker, W., Petković, M. (eds.) SDM 2010. LNCS, vol. 6358, pp. 150–168. Springer, Heidelberg (2010). doi:10.1007/978-3-642-15546-8_11

9. Soria-Comas, J., Domingo-Ferrer, J., Sánchez, D., Martínez, S.: Enhancing data utility in differential privacy via microaggregation-based k-anonymity. VLDB J. **23**(5), 771–794 (2014)

10. Sánchez, D., Domingo-Ferrer, J., Martínez, S., Soria-Comas, J.: Utility-preserving differentially private data releases via individual ranking microaggregation. Inf. Fusion **30**, 1–14 (2016)

11. Dwork, C.: Differential privacy. In: Bugliesi, M., Preneel, B., Sassone, V., Wegener, I. (eds.) ICALP 2006. LNCS, vol. 4052, pp. 1–12. Springer, Heidelberg (2006). doi:10.1007/11787006_1

12. Soria-Comas, J., Domingo-Ferrer, J.: Optimal data-independent noise for differential privacy. Inf. Sci. **250**, 200–214 (2013)

13. McSherry, F., Talwar, K.: Mechanism design via differential privacy. In: 48th Annual IEEE Symposium on Foundations of Computer Science-FOCS 2007, pp. 94–103. IEEE Computer Society, Washington D.C. (2007)

14. Nissim, K., Raskhodnikova, S., Smith, A.: Smooth sensitivity and sampling in private data analysis. In: 39th Annual ACM Symposium on Theory of Computing-STOC 2007, pp. 75–84. ACM, New York (2007)

15. Soria-Comas, J., Domingo-Ferrer, J., Sánchez, D., Martínez, S.: Improving the utility of differentially private data releases via k-anonymity. In: 12th IEEE International Conference on Trust, Security and Privacy in Computing and Communications-TrustCom 2013, pp. 372–379 (2013)

16. Sánchez, D., Domingo-Ferrer, J., Martínez, S.: Improving the utility of differential privacy via univariate microaggregation. In: Domingo-Ferrer, J. (ed.) PSD 2014. LNCS, vol. 8744, pp. 130–142. Springer, Cham (2014). doi:10.1007/978-3-319-11257-2_11

17. Domingo-Ferrer, J., Torra, V.: Ordinal, continuous and heterogeneous k-anonymity through microaggregation. Data Mining Knowl. Discov. **11**(2), 195–212 (2005)

18. Brand, R., Domingo-Ferrer, J., Mateo-Sanz, J.M.: Reference data sets to test and compare SDC methods for the protection of numerical microdata. Deliverable of the EU FP5 "CASC" project (2002). http://neon.vb.cbs.nl/casc/CASCtestsets.htm

19. Domingo-Ferrer, J., Sebé, F., Solanas, A.: A polynomial-time approximation to optimal multivariate microaggregation. Comput. Math. Appl. **55**(4), 714–732 (2008)

A Methodology to Compare Anonymization Methods Regarding Their Risk-Utility Trade-off

Josep Domingo-Ferrer, Sara Ricci$^{(\boxtimes)}$, and Jordi Soria-Comas

UNESCO Chair in Data Privacy, Department of Computer Science and Mathematics, Universitat Rovira i Virgili, Av. Països Catalans 26, 43007 Tarragona, Catalonia
{josep.domingo,sara.ricci,jordi.soria}@urv.cat

Abstract. We present here a methodology to compare statistical disclosure control methods for microdata in terms of how they perform regarding the risk-utility trade-off. Previous comparative studies (e.g. [3]) usually start by selecting some parameter values for a set of SDC methods and evaluate the disclosure risk and the information loss yielded by the methods for those parameterizations. In contrast, here we start by setting a certain risk level (resp. utility preservation level) and then we find which parameter values are needed to attain that risk (resp. utility) under different SDC methods; finally, once we have achieved an equivalent risk (resp. utility) level across methods, we evaluate the utility (resp. the risk) provided by each method, in order to rank methods according to their utility preservation (resp. disclosure protection), given a certain level of risk (resp. utility) and a certain original data set. The novelty of this comparison is not limited to the above-described methodology: we also justify and use general utility and risk measures that differ from those used in previous comparisons. Furthermore, we present experimental results of our methodology when used to compare the utility preservation of several methods given an equivalent level of risk for all of them.

Keywords: Record linkage · Disclosure risk · Utility preservation · Privacy · Permutation paradigm

1 Introduction

With the expansion of information technology, the importance of data analysis (e.g. to support decision making processes) has increased significantly. Although data collection has become easier and more affordable than ever before, releasing data for secondary use (that is, for a purpose other than the one that triggered the data collection) remains very important: in most cases, researchers cannot afford collecting themselves the data they need. However, when the data released for secondary use refer to individuals, households or companies, the privacy of the data subjects must be taken into account.

Statistical disclosure control (SDC) methods aim at releasing data that preserve their statistical validity while protecting the privacy of each data subject.

© Springer International Publishing AG 2017
V. Torra et al. (Eds.): MDAI 2017, LNAI 10571, pp. 132–143, 2017.
DOI: 10.1007/978-3-319-67422-3_12

Among the possible types of data releases, this work focuses on microdata (that is, on the release of data about individual subjects).

While there is a great diversity of SDC methods for microdata protection, all of them imply some level of data masking. The greater the amount of masking, the greater are both privacy protection and information loss. Different SDC methods tackle the trade-off between privacy and utility in different ways. For example, in global recoding the level of information loss is set beforehand (the amount of coarsening of the categories of each attribute), whereas the disclosure risk is evaluated afterwards on the protected data set. In contrast, in k-anonymity [9] the risk of disclosure (the risk of record re-identification, in particular) is set beforehand, whereas the actual information loss results from the masking needed to attain the desired level of disclosure risk.

Although some general assertions about specific SDC methods/models can be made, comparing the latter regarding the privacy-utility trade-off is not straightforward. Let us illustrate this point with two well-known privacy models: differential privacy [5] and k-anonymity [9]. In terms of privacy protection, ϵ-differential privacy is regarded as stronger than k-anonymity. On the contrary, k-anonymity is regarded as more utility-preserving than ϵ-differential privacy. The practical value of these general statements is dubious. After all, by increasing ϵ we reduce the protection of differential privacy, and by increasing k we reduce the utility of k-anonymous data. An accurate comparison between SDC methods has to take into consideration both aspects of the privacy-utility trade-off.

1.1 Contribution and Plan of this Paper

Many risk and utility measures have been proposed in the literature, but some of them are designed for use with specific SDC methods. For example, the probability of record re-identification is the natural risk measure in k-anonymity, but it may not be appropriate in SDC methods that are not predicated on protecting privacy by hiding each data subject within a crowd. In this work, we propose a framework based on general empirical measures of utility and risk to compare the risk-utility trade-off of several SDC methods.

Previous comparative studies (e.g. [3]) usually start by selecting some parameter values for a set of SDC methods and evaluate the disclosure risk and the information loss yielded by the methods for those parameterizations. In contrast, here we start by setting a certain risk level (or a certain utility level) and then we find which parameter values are needed to attain that risk (resp. that utility) under different SDC methods; finally, once we have achieved an equivalent risk level (resp. utility level) across methods, we evaluate the utility (resp. the risk) provided by each method, in order to rank methods according to their utility preservation (resp. disclosure protection), given a certain level of risk (resp. utility) and a certain original data set. Furthermore, we present experimental work that illustrates the application of the proposed methodology.

The rest of the paper is organized as follows. In Sect. 2, we introduce some background relevant to the remaining sections. In Sect. 3, we describe the proposed framework for comparing methods regarding their risk-utility trade-off.

In Sect. 4, we propose an empirical measure of disclosure risk that is based on record linkage. Experimental results are reported in Sect. 5. Conclusions are gathered in Sect. 6.

2   Background

2.1   Permutation Paradigm and Permutation Distance

In [2], a permutation paradigm to model anonymization was proposed. Let $X = \{x_1, \ldots, x_n\}$ be the values taken by attribute X in the original data set. Let $Y = \{y_1, \ldots, y_n\}$ represent the anonymized version of X. Consider the attribute Z obtained using the following reverse-mapping procedure

For $i = 1$ to n
    Compute $j = rank(y_i)$
    Set $z_i = x_{(j)}$ (where $x_{(j)}$ is the value of X of rank j)
Endfor

We can now view the anonymization of X into Y as a permutation step to turn X into Z, plus a small noise addition to turn Z into Y. Note the noise addition must be necessarily small, because it cannot alter ranks: by construction the ranks of Y and Z are the same. If we perform the above procedure independently for all attributes of an original data set **X** and corresponding attributes of an anonymized data set **Y**, we can say that anonymization can be decomposed into a permutation step to obtain a data set **Z** plus a (small) noise addition to obtain **Y** from **Z**.

The permutation distance measures the dissimilarity between two records in terms of the ranks of the values of their attributes. Assume the original data set **X** consists of m attributes $X^1, \ldots, X^m$ and the anonymized data set consists of corresponding attributes $Y^1, \ldots, Y^m$. Let $\mathbf{x} = (x^1, \ldots, x^m)$ be a record in **X** and $\mathbf{y} = (y^1, \ldots, y^m)$ be a record in **Y**. The permutation distance between **x** and **y** is the maximum of the rank distances of the attributes:

$$d(\mathbf{x}, \mathbf{y}) = \max_{1 \leq i \leq m} |rank(x^i) - rank(y^i)|.$$

The permutation distance between records is used in [2] to conduct a record linkage between the original data set **X** and the anonymized data set **Y**. In particular, records with minimal permutation distance are linked.

2.2   Utility Measures

Utility measures are a key component to compare SDC methods. We introduce two utility measures that will be used in the empirical evaluation of the proposed methodology: the propensity scores [12] and the earth mover's distance [8].

Algorithm 1 shows a way to use the propensity scores as a utility measure.

Algorithm 1

1. *Merge the original data set* **X** *and the anonymized data set* **Y** *and add a binary attribute* T *with value 1 for the anonymized records and 0 for the original records.*
2. *Regress* T *on the rest of attributes of the merged data set and call the adjusted attribute* $\hat{T}$. *Let the propensity score* $\hat{p}_i$ *of record* i *of the merged data set be the value of* $\hat{T}$ *for record* i.
3. *The utility can be considered high if the propensity scores of the anonymized and original records are similar. Hence, if the original and the anonymized data sets have the same number* n *of records, the following is a utility measure*

$$\mathcal{U}_{ps}(\mathbf{X}, \mathbf{Y}) = \frac{1}{2n} \sum_{i=1}^{2n} [\hat{p}_i - \frac{1}{2}]^2. \tag{1}$$

The value $\mathcal{U}_{ps}$ resulting from Eq. (1) is close to zero if the propensity scores computed with the regression model for all records are similar (in which case they will be neither 0 nor 1, but close to $1/2$). This situation means that the original and the anonymized records cannot be distinguished by the regression model, and hence the utility of the anonymized data set is high (its records "look" like the original records). In contrast, if the adjusted propensity scores were exactly the original values of T, it would mean that the regression model can exactly tell the original from the anonymized records, so the utility of the latter is low; in this case, we would have n propensity scores 0 and n propensity scores 1, which would yield a large $\mathcal{U}_{ps}$. Obviously, propensity scores as a utility measure are very dependent on the accuracy of the regression model adjusted to the data: the more accurate the model, the more discriminating it is and the less likely are values of $\mathcal{U}_{ps}$ indicating good utility (close to 0).

The earth mover's distance (EMD) is a natural extension of the notion of distance between single elements to distance between sets, or distributions, of elements. Given two distributions, one can be seen as a mass of earth in the space and the other as a collection of holes in that same space. Then, the EMD measures the least amount of work needed to fill the holes with earth, i.e. the minimal cost needed to transform one distribution into another by moving distribution mass. Thus, the EMD distance can be used to evaluate the similarity between the distribution of the original data set and the distribution of the anonymized data set. Note here that measuring similarity amounts to measuring utility, because, the more similar the distribution of the anonymized data to the distribution of the original data, the more useful are the anonymized data.

Formally, we can group records in clusters and represent each cluster j by its mean and the fraction ω_j of records that belong to that cluster. Let the original data set **X** be clustered as $\{(t_1, \omega_{t_1}), \dots, (t_h, \omega_{t_h})\}$, and the anonymized data set **Y** as $\{(q_1, \omega_{q_1}), \dots, (q_k, \omega_{q_k})\}$. Let $D = (d_{ij})$ be the matrix of the distance between the h clusters of **X** and the k clusters of **Y**, i.e. $d_{ij} = t_i - q_j$ (in the multivariate case, we take the Euclidean distance between cluster means).

The problem is to find a flow $F = (f_{ij})$, with f_{ij} being the flow between t_i and q_j, that minimizes the overall cost under some constraints (see [8] for more details). Once the optimal flow F is found, the earth mover's distance is defined as the resulting work normalized by the total flow:

$$\mathcal{U}_{emd}(\mathbf{X}, \mathbf{Y}) = \frac{\sum_{i=1}^{h} \sum_{j=1}^{k} d_{ij} f_{ij}}{\sum_{i=1}^{h} \sum_{j=1}^{k} f_{ij}} \tag{2}$$

The greater $\mathcal{U}_{emd}$ is, the more different are the distributions of $\mathbf{X}$ and $\mathbf{Y}$ and hence the more utility has been lost in the anonymization process.

3   A Methodology for Comparing the Risk-Utility Trade-off in SDC

In this section we describe a methodology for comparing SDC methods. Looking only at either the disclosure risk or the utility of an SDC method would be a flawed comparison. We need to analyze the privacy-utility trade-off, as explained in the introduction. Even if this principle may seem evident, very often it is not followed.

To make the proposed methodology as general as possible, we will employ empirical measures of risk and utility. That is, we will choose risk and utility measures that depend on the original and the anonymized data sets, rather than being prior conditions. To select specific measures, we need to define the aspects of risk and utility that we consider relevant for our comparison. In turn, the choice of measures will shape the outcome of the evaluation.

Let us illustrate the difference between empirical measures and prior conditions by taking differential privacy as an example. As a privacy model, differential privacy states some privacy guarantees but does not tell how they ought to be attained. Let us assume that A_1 and A_2 are ϵ-differentially private algorithms that output a data set. Let us also assume that A_2 is a refined version of A_1 that manages to attain ϵ-differential privacy while adding less noise than A_1. If we use the level ϵ of differential privacy as our risk measure, both A_1 and A_2 are equally good (they are both ϵ-differentially private). However, the fact that A_1 adds more noise to the original records may indicate that the data set output by A_1 entails less disclosure risk than the data set generated by A_2, even if differential privacy is unable to capture the difference. Alternative measures of disclosure risk (e.g. risk measures based on record linkage) should be able to capture the difference in risk between A_1 and A_2. In this work, we do not deny the value of any measure of disclosure risk, but, due to their broader applicability, we will employ empirical risk measures based on record linkage.

Let us assume that we are given functions

$$\mathcal{U} : \mathcal{D} \times \mathcal{D} \to \mathbb{R}$$

$$\mathcal{R} : \mathcal{D} \times \mathcal{D} \to \mathbb{R}$$

such that, for any given original data set $\mathbf{X}$ and anonymized data set $\mathbf{Y}$,

- $\mathcal{U}(\mathbf{X}, \mathbf{Y})$ measures the utility of $\mathbf{Y}$ as a replacement for $\mathbf{X}$.
- $\mathcal{R}(\mathbf{X}, \mathbf{Y})$ measures the disclosure risk of $\mathbf{Y}$ as a replacement for $\mathbf{X}$.

We have described some utility measures in Sect. 2.2. In Sect. 4 we will describe several risk measures based on record linkage.

SDC methods usually accept some parameters that can be adjusted to select the desired level of disclosure risk/utility. Let $M_\alpha(\mathbf{X})$ be the anonymized data output by SDC method M with parameter α when applied to data set $\mathbf{X}$.

Given an original data set $\mathbf{X}$ and two anonymization algorithms M^1 and M^2, we say that M^1 is more utility-preserving than M^2 at risk level r if

$$\mathcal{U}(\mathbf{X}, M_\alpha^1(\mathbf{X})) \geq \mathcal{U}(\mathbf{X}, M_\beta^2(\mathbf{X})),$$

for α and β such that $\mathcal{R}(\mathbf{X}, M_\alpha^1(\mathbf{X})) = \mathcal{R}(\mathbf{X}, M_\beta^2(\mathbf{X})) = r$.

In a similar fashion, we can compare the risk associated to a given level of data utility. We say that M^1 is less disclosive than M^2 at utility level u if

$$\mathcal{R}(\mathbf{X}, M_\alpha^1(\mathbf{X})) \leq \mathcal{R}(\mathbf{X}, M_\beta^2(\mathbf{X})),$$

for α and β such that $\mathcal{U}(\mathbf{X}, M_\alpha^1(\mathbf{X})) = \mathcal{U}(\mathbf{X}, M_\beta^2(\mathbf{X})) = u$.

The results of the previous utility (resp. risk) comparison depend not only on the SDC method, but also on the original data set, the risk and the utility measures selected, and the target level of risk (resp. utility). Actually, this comparison methodology is designed for use by a data controller who must decide which among several SDC methods is best suited to anonymize a given data set with a given target level of disclosure risk or utility. In other words, the aim is not to make general statements about the relative goodness of several SDC methods. Although such statements may make sense in some cases, our results can only be taken as empirical clues of such underlying truths.

4   Empirical Measures of Disclosure Risk

To compare the risk-utility trade-off between SDC methods, we need adequate measures of disclosure risk. For the methodology described in Sect. 3 to be broadly applicable, the risk measure should be as general as possible (rather than based on specific characteristics of an SDC method).

We propose a risk measure based on record linkage [11], which is a technique that seeks to match original records that correspond to the same individual. Among its several uses, record linkage has a direct application to disclosure risk assessment [10]. Such an application bears some resemblance to the way an intruder having access to the anonymized data and to some side knowledge would proceed. Let $\mathbf{E}$ be a data set that represents the non-anonymous side information available to the intruder. By linking records in $\mathbf{E}$ to records in $\mathbf{Y}$, the intruder associates identities to the records in $\mathbf{Y}$.

The number (or the proportion) of correct re-identifications is a common record linkage-based measure of disclosure risk. However, this measure has some limitations that we next discuss. It is certainly appropriate when SDC is achieved by masking the quasi-identifier attributes, whereas the sensitive attributes are left unmodified (or are only slightly modified). However, if the sensitive attributes have been significantly altered, a correct linkage may not be equivalent to disclosure. Furthermore, if we use SDC methods that are not based on masking the original records, we may not even be able to tell what a correct linkage is. Generating a synthetic data set by repeatedly sampling from a statistical model adjusted on the original data is an example of an SDC method not based on masking; and indeed, it is not possible to say what is the correct mapping between the original records and the synthetic records.

In the spirit of [2], rather than measuring the disclosure risk as the proportion of correct re-identifications, we will measure the risk of disclosure associated to a record in the original data set $\mathbf{X}$ by means of a distance to its linked record in $\mathbf{Y}$. Such an approach has two important advantages with respect to counting the number of correct re-identifications:

- It is more broadly applicable. The linkage between records in $\mathbf{X}$ and $\mathbf{Y}$ can be performed independently of the SDC methodology used, even when the correct mapping between original and anonymized records cannot be established.
- The distance between a record in $\mathbf{X}$ and its linked record in $\mathbf{Y}$ provides more detailed information about the risk associated to a record in $\mathbf{X}$ than a mere binary outcome (right/wrong linkage):
  - On the one hand, the binary nature of correct linkages could lead to understating the risk of disclosure when, in spite of failing to find the correct linkage, the intruder links to a record that is similar to the correct one.
  - On the other hand, if all the attributes in $\mathbf{Y}$ have been thoroughly altered by the SDC method, a correct linkage may not disclose any useful information to the intruder; in this case, the proportion of correct linkages would overstate the risk of disclosure.

Any record $\mathbf{x}$ in the original data set $\mathbf{X}$ is linked to the record $\mathbf{y_x} \in \mathbf{Y}$ at the smallest distance, that is, such that

$$d(\mathbf{x}, \mathbf{y_x}) = d(\mathbf{x}, \mathbf{Y}) = \min_{\mathbf{y} \in \mathbf{Y}} d(\mathbf{x}, \mathbf{y}).$$

The distance $d(\mathbf{x}, \mathbf{Y})$ is an indicator of the disclosure risk associated to $\mathbf{x}$. If the distance is small, there is a record in $\mathbf{Y}$ that is quite similar to $\mathbf{x}$ and the risk of disclosure is high.

The choice of the distance $d(\mathbf{x}, \mathbf{Y})$ is an important step in determining the disclosure risk. Along the lines of the permutation paradigm (see Sect. 2), our proposal is based on ranks, but it differs from [2] in the way attributes are aggregated. Let $\mathbf{x} = (x^1, \dots, x^m)$ be a record from an original data set $\mathbf{X}$ with attributes $X^1, \dots, X^m$ and $\mathbf{y} = (y^1, \dots, y^m)$ be a record from an anonymized data set $\mathbf{Y}$ with attributes $Y^1, \ldots, Y^m$. Take the distance between $\mathbf{x}$ and $\mathbf{y}$ to be the Euclidean distance between ranks, that is,

$$d(\mathbf{x}, \mathbf{y}) = \sqrt{\sum_{i=1}^{m} [rank_{X^i}(x^i) - rank_{Y^i}(y^i)]^2},$$

where the subscript of the rank function denotes the attribute within which the rank of the value in the argument is computed.

The overall risk of disclosure is an aggregation of the distances $d(\mathbf{x}, \mathbf{Y})$ for all $\mathbf{x} \in \mathbf{X}$. Many different aggregations are possible. In this work we focus on the average risk of disclosure by computing the mean of the record distances.

$$\mathcal{R}(\mathbf{X}, \mathbf{Y}) = \frac{1}{n} \log \sum_{\mathbf{x} \in \mathbf{X}} d(\mathbf{x}, \mathbf{Y}). \tag{3}$$

The smaller $\mathcal{R}(X, Y)$, the greater the risk of disclosure. The logarithm accounts for the fact that in disclosure risk the focus is on small distances. Without the logarithm, a large distance for a single record $\mathbf{x} \in \mathbf{X}$ could reduce in a significant manner the perception of risk for the overall data set; the logarithm reduces the influence of large distances.

5   Experimental Results

In this section we apply the methodology described in Sect. 3 to analyze the relative goodness of several anonymizations. Experiments are conducted by taking as original data the "Census" and "EIA" data sets [1], which are usual test sets in the SDC literature. The "Census" contains 13 numerical attributes and 1080 records, and "EIA" contains 11 numerical attributes and 4092 records.

The anonymized data sets have been generated by applying the following methods:

- *Correlated noise addition.* Multivariate normally distributed noise is added to the records in the collected data set, that is

$$\mathbf{Y} = \mathbf{X} + N(\mathbf{0}, \gamma \Sigma),$$

  where Σ is the covariance matrix of $\mathbf{X}$ and γ is an input parameter. Note that the covariance matrix of $\mathbf{Y}$ is proportional to the covariance matrix of $\mathbf{X}$.
- *Multiplicative noise.* We have used Höhne's variant ([6] and Ch. 3 of [7]). In a first step, each attribute value $x_j^i \in \mathbf{X}$ is multiplied by $1 \pm N(0, s)$, where s is an input parameter. Then, a transformation is applied to preserve the first and second-order moments.
- *Multivariate microaggregation.* We have used the MDAV heuristic [4]. In microaggregation, we partition the records of $\mathbf{X}$ in groups of k or more records, where records in a group are as similar as possible, and we replace each record by the corresponding centroid.

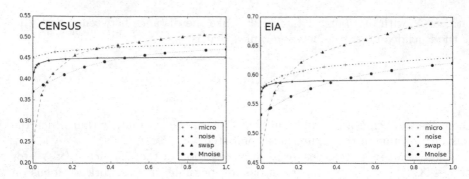

Fig. 1. Disclosure risk computed according to Eq. (3) for the anonymization methods under test and several input parameters. The x-axis shows the input parameter of the anonymization method (k, γ, p and s, respectively), so its scale should be disregarded. The y-axis shows the disclosure risk value. Left, "CENSUS" data set. Right, "EIA" data set.

- *Rank swapping.* Independently for each attribute, this method swaps the attribute's values within a restricted range: the ranks of two swapped values cannot differ by more than $p\%$ of the total number of records, where p is an input parameter.

More details about these methods can be found in [7].

5.1 Disclosure Risk Assessment

Recall that the comparison of anonymized data sets in Sect. 3 was performed on data sets that had either the same level of risk or the same level of utility. In this experimental work, we aim at determining which among the previous anonymization approaches gives better utility at a given level of disclosure risk. Thus, the first step is to find appropriate parameters for the previous anonymization algorithms that result in a given level of disclosure risk.

Figure 1 shows the disclosure risk computed according to Eq. (3) for the anonymization methods under test:

1. The curve labeled "micro" shows the risk of multivariate microaggregation for values of $k \in \{5, 10, 15, 20, 25, 50\}$.
2. The curve labeled "noise" shows the risk of correlated noise addition when $\gamma \in \{0.01, 0.025, 0.05, 0.075, 0.1, 0.25, 0.5, 1, 3\}$.
3. The curve labeled "swap" shows the risk of rank swapping when $p \in \{0.01, 0.05, 0.075, 0.1, 0.2, 0.3, 0.4, 0.5, 0.6, 0.7, 0.8, 0.9\}$.
4. The curve labeled "Mnoise" shows the risk of multiplicative noise when $s \in \{0.05, 0.1, 0.2, 0.3, 0.4, 0.5, 0.6, 0.7, 0.9, 1\}$.

Table 1. Utility loss measured using propensity scores (Eq. (1)) and the earth mover's distance (Eq. (2)) for the anonymization methods under test and for input parameters that were found to yield the same level of disclosure risk.

Methods	CENSUS		EIA	
	Propensity	EMD	Propensity	EMD
Microaggregation	4.28×10^{-4}	0.16	2.17×10^{-5}	0.040
Correlated noise addition	3.83×10^{-2}	0.38	4.22×10^{-5}	0.065
Rank swapping	3.51×10^{-3}	0.28	9.01×10^{-4}	0.091
Multiplicative noise	6.3×10^{-3}	0.29	9.85×10^{-5}	0.066

For the "Census" data set, a possible match between methods occurs at $\mathcal{R}(X, Y) = 0.45$ and is given by:

1. multivariate microaggregation with $k = 5$,
2. correlated noise addition with $\gamma = 1$,
3. rank swapping with $p = 0.2$, and
4. multiplicative noise with $s = 0.5$.

The microaggregation cluster size $k = 5$ may seem small compared to the parameter values that we get for the other methods. However, such a difference in magnitude can be explained by the fact that multivariate microaggregation is known to yield poorly homogeneous clusters when the number of dimensions is large, even if the cluster size k is small.

For the "EIA" data set, a possible match between methods occurs at $\mathcal{R}(X, Y) = 0.58$ and is given by:

1. multivariate microaggregation with $k = 5$,
2. correlated noise addition with $\gamma = 0.05$,
3. rank swapping with $p = 0.08$, and
4. multiplicative noise with $s = 0.3$.

5.2 Utility Assessment

We evaluate the utility of the anonymization methods for the parameters above that were found to yield the same level of disclosure risk. The utility is evaluated using the measures based on propensity scores and EMD, that were described in Sect. 2.2.

We found in Sect. 5.1 that, for the "Census" data set, the SDC methods being compared with parameters $k = 5, \gamma = 1, p = 0.2$ and $s = 0.5$, respectively, yielded the same risk of disclosure. By comparing the utility measures for these methods, we can determine which among them is preferable in this case. Table 1 shows the results for the propensity scores and EMD measures. Both utility measures are consistent and tell us that microaggregation has the best utility, followed by rank swapping, multiplicative noise and, finally, correlated noise addition.

For the "EIA" data set, the SDC methods being compared with parameters $k = 5$, $\gamma = 0.05$, $p = 0.08$ and $s = 0.3$, respectively, yielded the same risk of disclosure. The utility results for the propensity scores and EMD measures for this data set are shown in Table 1. Like in the other data set, methods are consistently ranked by the both measures, but the ranking is different: multivariate microaggregation has the best utility, followed by correlated noise addition, multiplicative noise, and, finally, rank swapping.

The results have shown that the SDC methods under comparison perform differently in different situations. Multivariate microaggregation always had the best utility (at the given level of disclosure risk), but the relative utility performance of the other methods changed between "Census" and "EIA". This shows that, unless there are good reasons for using a given anonymization method, it is usually better to make several anonymizations at the desired level of disclosure risk and select the one that has the greatest utility.

6   Conclusions

We have described a methodology to compare different anonymizations in terms of the risk-utility trade-off they attain. It is not enough to compare methods based on the level of risk or the utility they provide, because that gives only a partial picture.

We have proposed a disclosure risk measure based on record linkage and in the spirit of the permutation paradigm (which tells that disclosure risk control comes essentially from rank permutation)

We have contributed an experimental analysis for two well-known data sets and four well-known anonymization methods. The results differ between data sets. As a conclusion from the experimental analysis, the best strategy seems to be to make several anonymizations at the desired level of disclosure risk and select the one that has the greatest utility.

Acknowledgments and Disclaimer. The following funding sources are gratefully acknowledged: European Commission (projects H2020 644024 "CLARUS" and H2020 700540 "CANVAS"), Government of Catalonia (ICREA Acadèmia Prize to J. Domingo- Ferrer and grant 2014 SGR 537), Spanish Government (projects TIN2011-27076-C03-01 "CO-PRIVACY", TIN2014-57364-C2-R "SmartGlacis" and TIN2016-80250-R, "Sec-MCloud"). The authors are with the UNESCO Chair in Data Privacy, but the views in this paper are their own and do not necessarily reflect those of UNESCO.

References

1. Brand, R., Domingo-Ferrer, J., Mateo-Sanz, J.M.: Reference data sets to test and compare SDC methods for protection of numerical microdata. European Project IST-2000-25069 CASC (2002)

2. Domingo-Ferrer, J., Muralidhar, K.: New directions in anonymization: permutation paradigm, verifiability by subjects and intruders, transparency to users. Inf. Sci. **337**, 11–24 (2016)
3. Domingo-Ferrer, J., Torra, V.: A quantitative comparison of disclosure control methods for microdata. In: Doyle, P., Lane, J.I., Theeuwes, J.J.M., Zayatz, L. (eds.) Confidentiality, Disclosure and Data Access: Theory and Practical Applications for Statistical Agencies, pp. 111–134. North-Holland, Amsterdam (2001)
4. Domingo-Ferrer, J., Torra, V.: Ordinal, continuous and heterogeneous k-anonymity through microaggregation. Data Mining Knowl. Discov. **11**(2), 195–212 (2005)
5. Dwork, C.: Differential privacy. In: Bugliesi, M., Preneel, B., Sassone, V., Wegener, I. (eds.) ICALP 2006. LNCS, vol. 4052, pp. 1–12. Springer, Heidelberg (2006). doi:10.1007/11787006_1
6. Höhne, J.: Varianten von Zufallsüberlagerung (in German). Working paper of the project "Faktische Anonymisierung wirtschaftsstatistischer Einzeldaten" (2004)
7. Hundepool, A., Domingo-Ferrer, J., Franconi, L., Giessing, S., Nordholt, E.S., Spicer, K., De Wolf, P.P.: Statistical Disclosure Control. Wiley, Chichester (2012)
8. Rubner, Y., Tomasi, C., Guibas, L.J.: The earth mover's distance as a metric for image retrieval. Int. J. Comput. Vis. **40**(2), 99–121 (2000)
9. Samarati, P., Sweeney, L.: Protecting privacy when disclosing information: k-anonymity and its enforcement through generalization and suppression. Technical report, SRI International (1998)
10. Torra, V., Domingo-Ferrer, J.: Record linkage methods for multidatabase data mining. In: Torra, V. (ed.) Information Fusion in Data Mining, pp. 101–132. Springer, Heidelberg (2003). doi:10.1007/978-3-540-36519-8_7
11. Winkler, W.E.: Matching and Record Linkage. Wiley, New York (1995)
12. Woo, M.J., Reiter, J.P., Oganian, A., Karr, A.F.: Global measures of data utility for microdata masked for disclosure limitation. J. Priv. Confidentiality **1**(1), 7 (2009)

Session-Based Network Intrusion Detection Using a Deep Learning Architecture

Yang Yu, Jun Long, and Zhiping Cai[✉]

School of Computer, National University of Defense Technology,
Changsha 410073, Hunan, China
yuyang_nudt@163.com, {junlong,zpcai}@nudt.edu.cn

Abstract. Intrusion detection is extremely crucial to prevent computer systems from being compromised. However, as numerous complicated attack types have growingly appeared and evolved in recent years, obtaining quite high detection rates is increasingly difficult. Also, traditional heavily hand-crafted evaluation datasets for network intrusion detection have not been practical. In addition, deep learning techniques have shown extraordinary capabilities in various application fields. The primary goal of this research is utilizing unsupervised deep learning techniques to automatically learn essential features from raw network traffics and achieve quite high detection accuracy. In this paper, we propose a session-based network intrusion detection model using a deep learning architecture. Comparative experiments demonstrate that the proposed model can achieve incredibly high performance to detect botnet network traffics.

Keywords: Network intrusion detection · Deep learning · Raw network traffics · Session

1 Introduction

Network intrusion detection systems (NIDSs) play increasingly significant roles in protecting computer systems from malicious attacks. Traditionally, NIDSs are divided into signature-based NIDSs and anomaly-based NIDSs. A signature-based NIDS detects known attacks through matching established rules and patterns. In contrast, an anomaly-based NIDS involved in unknown attacks detection identifies attacks via discovering deviations from normal activities. The growing increase of new attacks and large volumes of network traffic data are posing incredible challenges to the information industry [3,4]. In addition, deep learning techniques have been successfully applied in numerous research fields (e.g., image recognition). Therefore, deep learning approaches are expected to obtain rather impressive performance in anomaly detection.

Existing deep learning approaches utilized in intrusion detection are categorized into unsupervised deep learning methods and supervised deep learning methods. The difference between supervised and unsupervised methods is whether

© Springer International Publishing AG 2017
V. Torra et al. (Eds.): MDAI 2017, LNAI 10571, pp. 144–155, 2017.
DOI: 10.1007/978-3-319-67422-3_13

training data are labeled. Specifically, unsupervised deep learning methods using unlabeled data include sparse auto-encoder [11], restricted boltzmann machine (RBM) [6], deep belief network (DBN) [9, 21] and recurrent neural network (RNN) [8]. The supervised deep learning methods include convolutional neural network (CNN) which combines with multi-layer perceptron (MLP) in [19]. Moreover, stacked autoencoders (SAE) algorithm obtained remarkable performance in traffic identification [17]. Machine learning methods combining with different deep learning approaches are also applied in intrusion detection, such as extreme learning machine (ELM) [18] and support vector machine (SVM) [5, 13]. However, the majority of these studies have been limited to evaluate their research on the outdated dataset (i.e., KDD Cup 99 dataset [10, 15] and NSL-KDD dataset [12]). In general, substantive network traffic data collected from the real world are always high dimensional and unlabeled. Consequently, unsupervised deep learning algorithms are usually utilized for dimensionality reduction and feature extraction.

As numerous novel attacks have increasingly appeared and evolved in recent years, traditional network intrusion detection techniques are facing several challenges. First, network intrusion techniques producing too many false positives and false negatives can not satisfy the requirement of high accuracy of NIDSs [14]. Second, intrusion features used for traditional machine learning methods are heavily structured and have special semantics involved in specific expert knowledge, which is extremely expensive and time-consuming. Finally, the heavily hand-crafted features are closely related to specific attack types. In other words, those features would fail to detect novel and complicated attacks. On the other hand, deep learning approaches can automatically learn essential features from large volumes of data. Moreover, unsupervised deep learning techniques are more potential than traditional machine learning methods for intrusion detection [20] due to the difficulty of obtaining labeled data.

The objective of our paper is utilizing unsupervised deep learning methods to automatically learn essential features from raw network packets and achieve quite high detection accuracy. In this study, we propose a session-based network intrusion detection model using a deep learning architecture. Comparative experiments demonstrate that the proposed model can achieve incredibly high performance for network intrusion detection. We obtain quite impressive performance through applying stacked denoising autoencoders (SDA) based deep learning architecture to detect botnet traffics.

The remainder of this paper is organized as follows. Section 2 describes a session-based network intrusion detection model using a deep learning architecture. Section 3 shows experimental results and discusses their specific meanings and implications. Finally, Sect. 4 concludes the study with some future work.

2   Methodology

In this paper, we propose a session-based network intrusion detection model using a deep learning architecture illustrated in Fig. 1. Generally, this model is composed of session-based data preprocessing module and a deep learning architecture module. Network session data can well reflect the correlation of network packets. Specifically, the session-based data preprocessing module extracts few simple features from packets' header portion and selects payloads of the network application layer within a session as features. The header features and payloads within a session together form a record. Subsequently, those records are fed into an unsupervised deep learning algorithm to obtain essential features for classifying normal and malicious network traffics.

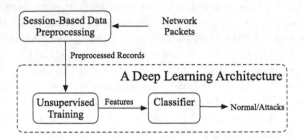

Fig. 1. The session-based network intrusion detection model using a deep learning architecture

2.1   Session-Based Data Preprocessing Method

A record is a training sample which consists of little packets' header information and raw payload data of the network application layer. There are three steps to transform raw network traffics into normalized records: session construction, record construction, and normalization. The data preprocessing procedure is presented in Fig. 2.

Session Construction. TCP, UDP, and ICMP packets are first utilized to construct sessions, respectively. The TCP, UDP, and ICMP sessions are separately defined by a five tuple which is referred to as a session ID which can identify a unique session. There is a one-to-one correspondence between the session ID and the record. Specifically, a TCP session ID, as well as a UDP session ID, consists of protocol type, IP source address, IP destination address, source port, and destination port. Similarly, an ICMP session ID is composed of protocol type, IP source address, IP destination address, ICMP type, and ICMP code. The payloads of the application layer within each session are separately joined.

Record Construction. Part of packets' header information and payloads within a session are extracted as features of a record. The dimensional distribution of the record features is described in Table 1. The length of each record is 1000. The first 17 positions of the record are reserved for the features of packets' header. The rest of positions is payloads of packets in a session. If the length of a record is less than 1000, zeroes are padded. Correspondingly, the extra part is truncated. The protocol type (TCP, UDP, ICMP) and time features of the IP header are selected to be features. The protocol type is represented as 100, 010,

Table 1. Dimensional distribution of the record features in our dataset

Feature name	Description	Length
protocol type	110 if protocol type is tcp; 010 if protocol type is udp; 001 if protocol type is icmp.	3
port	Source port and destination port	2
icmp_type	Value of type field of icmp packet	1
icmp_code	Value of code field of icmp packet	1
tcp_flags	Separate cumulative sum of FIN, SYN, RST, PSH, ACK, URG, ECE, CWR within a session in each dimension	8
interval_mean	Mean value of time intervals of all packets within a session	1
interval_variance	Variance value of time intervals of all packets within a session	1
payload	Partial payloads of the network application layer in a session	983

001 for TCP, UDP and ICMP, respectively. We eliminate source and destination IP addresses because they can be used to identify botnet traffics [1]. The time intervals of packets in a session are calculated, and their mean and variance values are used as time features of a record. For the TCP header, besides source and destination ports, numbers of TCP flags (FIN, SYN, RST, PSH, ACK, URG, ECE, CWR) of packets in a session are counted for each flag. For example, if there are only ten packets whose SYN flag is set to 1 in a session, the value of SYN feature would be 10.

Normalization. The payloads based on bytes are expressed as integer numbers between 0 to 255. The integer numbers then are normalized to float numbers from 0 to 1. For some features of the packet header, which we can not know numerical bounds, min-max normalization method is used to normalize each single feature to the range $[0, 1]$. The formula of the min-max normalization is $x' = \frac{x - \min(F)}{\max(F) - \min(F)}$, where $\min(F) = $ minimum value of feature F and $\max(F) = $ maximum value of feature F.

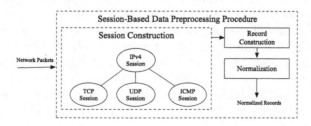

Fig. 2. The session-based data preprocessing procedure

2.2   Deep Learning Architecture Description

In this section, we present a SDA-based deep learning architecture for session-based network intrusion detection model. Stacked denoising autoencoders (SDA) [16] are the extension of classical stacked autoencoders. The structure and relevant terminology of a denoising autoencoder are first briefly introduced. Suppose there is an unlabeled training dataset $D = \left\{ x^{(1)}, x^{(2)}, ..., x^{(m)} \right\}$ with m samples, where the input sample $x^{(i)} \in \Re^n$ is an n-dimensional vector. The input of a denoising autoencoder is first stochastically corrupted via a stochastic mapping $\tilde{x} \sim q_D\left(\tilde{x} | x \right)$. In particular, the corruption approach in our experiments is to randomly set some of the input units to zero. A denoising autoencoder then maps the corrupted input vector $\tilde{x}$ to a hidden representation h called code via a deterministic mapping:

$$h = f\left(W\tilde{x} + b \right), \tag{1}$$

where W is a weight matrix and b is a bias vector, and the mapping $f\left(\cdot \right)$ called the encoder is a sigmoid function (i.e., $f\left(z \right) = \frac{1}{1+e^{-z}}$) in our model. The code h then is transformed back into an n-dimensional vector $x = \hat{x}$ called the reconstruction of input x. The transformation is implemented by the same mapping called decoder:

$$\hat{x} = f\left(W'h + b' \right), \tag{2}$$

where we set the weight matrix W' to tied weighs which means W' is the matrix transpose of W (i.e., $W' = W^T$) in our experiments. The denoising autoencoder attempts to reconstruct the raw input from corrupted version of the input. In order to minimize the reconstruction error of the input and the output, the loss function in our experiments is the cross-entropy loss:

$$L\left(x, \hat{x} \right) = -\sum_{j=1}^{n} \left[x_j \log \hat{x}_j + \left(1 - x_j \right) \log \left(1 - \hat{x}_j \right) \right]. \tag{3}$$

Figure 3 depicts the specific steps that how to train a SDA-based deep learning architecture. The procedure of stacking denoising autoencoders is mainly divided into two stages. The first stage, unsupervised layer-wise pre-training stage, is a greedy layer-wise training process. Specifically, a denoising autoencoder neural network is first trained through minimizing the reconstruction error of the input and out. The second denoising autoencoder is then trained by taking the hidden-layer output of the fist autoencoder as input. Thus, the denoising autoencoders are stacked into a deep neural network through training a number of denoising autoencoders. The training set is utilized in the unsupervised layer-wise training stage only involved in unlabeled data. The second stage, supervised fine-tuning stage, adds a logistic regression layer for classification on the top of the stacked denoising autoencoders. In other words, the output of the last hidden layer of stacked autoencoders is the input of a softmax classifier. The entire neural network is then trained as a multilayer perceptron and optimize all the parameters using labeled samples. The validation set is used to select the best model, and early-stopping method is applied to avoid overfitting. Finally, the

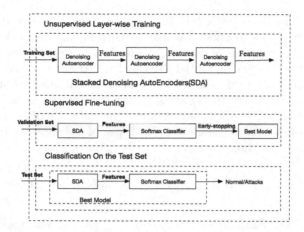

Fig. 3. Specific steps to train a SDA-based deep learning architecture

test set is employed for the classification task and evaluating the performance of classification results.

The SDA-based deep learning architecture has some advantages as follows. First, it can use numerous unlabeled samples to automatically learn important features. Second, the corruption strategy makes it obtain more robust features from missing data and noisy input. Finally, it provides the more effective method of dimensionality reduction than principal components analysis (PCA) when the hidden layer is non-linear [7].

3   Experimental Results and Performance Analysis

Our experiments included two types of classification tasks, namely, binary classification and multi-classification. Specifically, the binary classification task included normal and attacks (bots). The multi-classification task included eight types of classes, i.e., normal and seven types of bots. Two sizes of dataset were separately used to evaluate the performance of the SDA-based neural network architecture, and we compared this architecture with other state-of-art deep learning methods.

In this section, we first present the evaluation dataset and algorithmic parameter settings information. Next, we briefly introduce evaluation metrics for the performance analysis. Finally, we discuss experimental results of three-part experiments.

3.1   Dataset and Parameter Settings

There are two types of network traffics in our dataset, namely normal traffics and botnet traffics. The normal portion of our dataset is supplied by the UNB ISCX IDS 2012 dataset [2] because its data do not be heavily anonymized and have no privacy issues. The UNB ISCX IDS 2012 dataset includes seven days of normal and malicious network traffics. We select the first-day network traces which contain full packet payloads of normal activity in a pcap file format. In addition, We choose botnet traffics of CTU-13 dataset [1] to express cases of malicious behaviors in the real world. The CTU-13 dataset contains 13 botnet scenarios in which specific malwares (bots) are separately executed. Table 2 presents the distribution of traffic records in our dataset. The data size of the 43% dataset is 43% of the whole dataset. There are seven types of bot network traffics in both 43% dataset and the whole dataset.

Table 2. Distribution of traffic records in our dataset

43% dataset					Whole dataset				
Traffic		Training	Validation	Test	Total	Traffic		Training	Validation	Test	Total
Normal		18467	3700	3743	25910	Normal		41480	8123	8174	57867
Bot	Neris	3516	675	677	4868	Bot	Neris	8039	1567	1565	11171
	Rbot	6015	1185	1158	8358		Rbot	6073	1228	1221	8522
	Virut	507	102	114	723		Virut	18914	3680	3767	26361
	Menti	220	42	38	300		Menti	217	40	43	300
	Sogou	31	5	8	44		Sogou	34	5	5	44
	Murlo	1911	424	400	2735		Murlo	2013	364	358	2735
	NSIS.ay	4436	867	862	6165		NSIS.ay	4395	903	867	6165
	Total	16636	3300	3257	23193		Total	39685	7877	7826	55298
Total		35103	7000	7000	49103	Total		81165	16000	16000	113165

The 43% dataset and the whole dataset are used to evaluate the performance of the SDA-based neural network architecture, respectively. The proportion of the training, validation and test set is 5:1:1. There were three hidden layers in the SDA neural network. The number of hidden units was all simply set to 1000 and the corruption level for training each denoising autoencoder was separately set to 10%, 20%, and 30%. The corruption level in our experiments means how many input units of a denoising autoencoder are randomly set to 0.

3.2   Evaluation Metrics

Generally, the performance of network intrusion detection is evaluated through four metrics, i.e., accuracy (AC), precision (P), recall (R), F-measure (F). These metrics are further defined by true positives (TP), true negatives (TN), false positives (FP) and false negatives (FN). They can be obtained by a confusion matrix C whose element $C_{i,j}$ equals to the number of samples of class i predicted to class j. Specifically, TP and TN are the numbers of attack and normal records predicted correctly, respectively. Accordingly, FP and FN are separately the number of normal and attack records classified incorrectly.

- Accuracy (AC): presents the percentage of true prediction over all records.

$$AC = (TP + TN) / (TP + TN + FP + FN) \qquad (4)$$

- Precision (P): presents the ability of a classifier predicting actual attacks over all attack records it predicted.

$$P = TP/(TP + FP) \tag{5}$$

- Recall (R): presents the percentage of predicted attack records versus all the attack records.

$$R = TP/(TP + FN) \tag{6}$$

- F-measure (F): can better evaluate the performance due to the combination of recall and precision.

$$F = 2PR/(P + R) \tag{7}$$

In addition, receiver operating characteristic (ROC) metric can evaluate the result quality of the classifier through false positive rate and true positive rate. The left top corner of a ROC figure means the highest true positive rate and the lowest false positive rate. Hence, the closer that the value of the area under the ROC curve (AUC) is to 1, the better result is.

3.3   Experimental Results

We discuss results of three-part experiments: binary classification using the SDA-based deep neural network, multi-classification using the SDA-based deep neural network, and classification using different deep learning architectures.

Binary Classification Using the SDA-Based Deep Neural Network. The results of binary classification are presented in Table 3. We found that the SDA approach obtained a little higher overall performance on the larger size dataset. This result suggests that the amounts of training data could influence the performance of the SDA approach and the SDA approach presents better learning ability with more training data. Also, we noticed that the values of all the evaluation metrics of the SDA approach achieved were over 99%. These results implicate that the SDA approach could learn highly significant features from the raw payloads of the network application layer.

Table 3. Metrics for binary classification

Dataset	Type	Accuracy (%)	Precision (%)	Recall (%)	F-measure (%)
43% dataset	Total	99.41	99.29	99.29	99.29
	Normal	99.25	99.25	99.41	99.33
	Attack	99.60	99.32	99.14	99.23
Whole dataset	Total	99.48	99.39	99.39	99.39
	Normal	99.46	99.45	99.36	99.41
	Attack	99.50	99.34	99.43	99.38

Additionally, the ROC curves for the two sizes of the dataset are shown in Fig. 4. As expected, AUC values of two experiments are high and equal. This result indicates that session-based network intrusion detection can achieve quite high true positive rate and false positive rate.

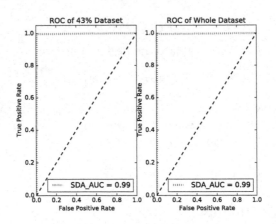

Fig. 4. ROC curves of the SDA for the binary classification task

Multi-classification Using the SDA-Based Deep Neural Network. As observed from Fig. 5a, the SDA approach performs well too in the multi-classification (8 class) task comparing with binary classification task on both the 43% dataset and the whole dataset. Specifically, the SDA approach achieved 98.11% accuracy rate on the whole dataset, whereas the SDA approach achieved 97.96% accuracy rate on the 43% dataset. In other words, the SDA approach performs better in the larger dataset for the multi-classification task, which is the same as the binary classification. We also observed that the precision, recall and F-measure values of two experiments were equal to corresponding accuracy value of each experiment. We found that the SDA approach achieved lesser metric values for the multi-classification task than the binary classification task. This is because sample sizes of some bots (e.g., Sogou and Menti) are too small to learn important features for the SDA approach in the multi-classification task, which would decrease the overall accuracy rate. Figure 5b displays the ROC curves for the multi-classification task. The AUC values of the ROC curves just are the same as binary classification. We obtained high AUC values in the multi-classification task too, which suggests that the SDA approach could perform well in the multi-classification task and learn robust features even when sample sizes of some classes are small.

Classification Using Different Deep Learning Architectures. We also evaluated some different deep learning approaches, namely SAE, DBN, and AE-CNN. Those methods have similar training steps described in Fig. 3. In the AE-CNN

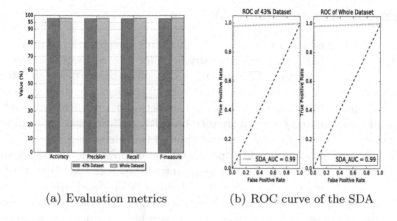

(a) Evaluation metrics   (b) ROC curve of the SDA

Fig. 5. Metrics and ROC curve for the multi-classification task

experiment, the raw input data of 1000 dimensions were first mapped to 784 dimensions through an autoencoder to simulate the image classification. CNN then took the compressed data as input. The number of hidden layers of these approaches was three. The classification accuracy of different deep learning approaches is shown in Table 4. Compared with other deep learning approaches, SDA achieved better overall performance in all experiments except multi-classification on the 43% dataset. Furthermore, while other approaches got worse performance when the dataset size increased, the SDA method achieved the highest accuracy on the whole dataset. The SDA approach also had the best performance in the binary classification on the two datasets. While SDA and DBN have the similar training principles, SDA appears to be comparable and superior to DBN. And the AE-CNN method always yields worse performance than others. These results suggest that SDA with the denoising criterion can learn significant higher level representation (features) from raw traffic data, and deep learning approaches have remarkable capabilities for the intrusion detection task.

Table 4. Classification accuracy of different deep learning approaches

Type	Dataset	SDA	SAE	DBN	AE-CNN
2-class	43% dataset	99.41%	99.26%	99.29%	98.46%
	Whole dataset	99.48%	99.42%	99.39%	98.54%
8-class	43% dataset	97.96%	98.51%	98.04%	96.37%
	Whole dataset	98.11%	97.96%	97.55%	93.58%

4   Conclusion and Future Work

The numerous derived features contained in existing benchmark intrusion detection datasets have not been practical anymore for the real-world and sophisticated intrusion detection. We implemented a SDA-based deep architecture automatically learn essential features of botnet traffics. We also proposed a session-based method to construct intrusion detection dataset from raw network traffics, and we evaluated the performance of different deep learning approaches on the dataset. The experimental results showed that deep learning approaches had high effectiveness and incredible potential to be applied in the session-based network intrusion detection. In future, we plan to adjust some parameters of deep learning architectures (e.g., the number of hidden layers) and utilize the entire payloads of the network application layer to further improve performance. In addition, the different hybrid approaches of combining machine learning and deep learning approaches for intrusion detection can also be promising and potential.

Acknowledgements. This work is supported by the National Natural Science Foundation of China under Grant Nos. 61379145, 61105050.

References

1. The CTU-13 dataset. https://stratosphereips.org/category/dataset.html
2. The UNB ISCX 2012 intrusion detection evaluation dataset. http://www.unb.ca/cic/research/datasets/ids.html
3. Cai, Z., Wang, Z., Zheng, K., Cao, J.: A distributed TCAM coprocessor architecture for integrated longest prefix matching, policy filtering, and content filtering. IEEE Trans. Comput. **62**(3), 417–427 (2013)
4. Cheng, J., Yin, J., Liu, Y., Cai, Z., Wu, C.: DDoS attack detection using IP address feature interaction. In: International Conference on Intelligent Networking and Collaborative Systems, INCOS 2009, pp. 113–118. IEEE (2009)
5. Erfani, S.M., Rajasegarar, S., Karunasekera, S., Leckie, C.: High-dimensional and large-scale anomaly detection using a linear one-class svm with deep learning. Pattern Recogn. **58**, 121–134 (2016)
6. Fiore, U., Palmieri, F., Castiglione, A., De Santis, A.: Network anomaly detection with the restricted Boltzmann machine. Neurocomputing **122**, 13–23 (2013)
7. Hinton, G.E., Salakhutdinov, R.R.: Reducing the dimensionality of data with neural networks. Science **313**(5786), 504–507 (2006)
8. Jung, W., Kim, S., Choi, S.: Poster: deep learning for zero-day flash malware detection. In: 36th IEEE Symposium on Security and Privacy (2015)
9. Li, Y., Ma, R., Jiao, R.: A hybrid malicious code detection method based on deep learning. Int. J. Secur. Appl. Methods **9**(5), 205–216 (2015)
10. McHugh, J.: Testing intrusion detection systems: a critique of the 1998 and 1999 DARPA intrusion detection system evaluations as performed by Lincoln laboratory. ACM Trans. Inf. Syst. Secur. (TISSEC) **3**(4), 262–294 (2000)
11. Niyaz, Q., Sun, W., Javaid, A.Y., Alam, M.: A deep learning approach for network intrusion detection system. In: Proceedings of the 9th EAI International Conference on Bio-inspired Information and Communications Technologies (Formerly BIONETICS), BICT, vol. 15, pp. 21–26 (2016)

12. Revathi, S., Malathi, A.: A detailed analysis on NSL-KDD dataset using various machine learning techniques for intrusion detection. Int. J. Eng. Res. Technol. (2013). ESRSA Publications
13. Salama, M.A., Eid, H.F., Ramadan, R.A., Darwish, A., Hassanien, A.E.: Hybrid intelligent intrusion detection scheme. In: Gaspar-Cunha, A., Takahashi, R., Schaefer, G., Costa, L. (eds.) Soft Computing in Industrial Applications, pp. 293–303. Springer, Heidelberg (2011). doi:10.1007/978-3-642-20505-7_26
14. Sommer, R., Paxson, V.: Outside the closed world: on using machine learning for network intrusion detection. In: 2010 IEEE Symposium on Security and Privacy (SP), pp. 305–316. IEEE (2010)
15. Tavallaee, M., Bagheri, E., Lu, W., Ghorbani, A.A.: A detailed analysis of the KDD CUP 99 data set. In: IEEE Symposium on Computational Intelligence for Security and Defense Applications, CISDA 2009, pp. 1–6. IEEE (2009)
16. Vincent, P., Larochelle, H., Lajoie, I., Bengio, Y., Manzagol, P.A.: Stacked denoising autoencoders: learning useful representations in a deep network with a local denoising criterion. J. Mach. Learn. Res. **11**((Dec)), 3371–3408 (2010)
17. Wang, Z.: The applications of deep learning on traffic identification. BlackHat USA (2015)
18. Yan, W., Yu, L.: On accurate and reliable anomaly detection for gas turbine combustors: a deep learning approach. In: Proceedings of the Annual Conference of the Prognostics and Health Management Society (2015)
19. Yao, Y., Wei, Y., Gao, F.x., Yu, G.: Anomaly intrusion detection approach using hybrid MLP/CNN neural network. In: Sixth International Conference on Intelligent Systems Design and Applications, ISDA 2006, vol. 2, pp. 1095–1102. IEEE (2006)
20. Yu, Y., Long, J., Liu, F., Cai, Z.: Machine learning combining with visualization for intrusion detection: a survey. In: Torra, V., Narukawa, Y., Navarro-Arribas, G., Yañez, C. (eds.) MDAI 2016. LNCS, vol. 9880, pp. 239–249. Springer, Cham (2016). doi:10.1007/978-3-319-45656-0_20
21. Yuan, Z., Lu, Y., Xue, Y.: Droiddetector: android malware characterization and detection using deep learning. Tsinghua Sci. Technol. **21**(1), 114–123 (2016)

Data Mining and Applications

A Hybrid Model of ARIMA and ANN with Discrete Wavelet Transform for Time Series Forecasting

Warut Pannakkong[✉] and Van-Nam Huynh

School of Knowledge Science, Japan Advanced Institute of Science and Technology,
Nomi, Ishikawa, Japan
{warut,huynh}@jaist.ac.jp

Abstract. Improving the prediction performance is a main objective in time series forecasting research area. Wavelet transform has been used for decomposing time series into approximation and detail before further analysis with forecasting models. However, generally, the approximation and the detail are assumed as either linear or nonlinear. In fact, the wavelet transform is not for decomposing the original time series into linear and nonlinear time series. Therefore, this study proposes a hybrid forecasting model of discrete wavelet transform (DWT), autoregressive integrated moving average (ARIMA) and artificial neural network (ANN) without linear or nonlinear assumption on the approximation and the detail. The proposed model decomposes the time series by the DWT to get the approximation and the detail. Then, the approximation and the detail are separately analyzed by Zhang's hybrid model involving the ARIMA and the ANN in order to capture both linear and nonlinear components of the approximation and the detail. Finally, the linear and nonlinear components are combined for final forecasting. The proposed model has been tested with three well-known data sets: Wolf's sunspot, Canadian lynx and British pound/US dollar exchange rate. The experimental results indicate that the proposed model can outperform the ARIMA, the ANN, and the Zhang's hybrid model in all three tested time series and measures (i.e. MSE, MAE and MAPE).

Keywords: Hybrid model · Time series forecasting · Autoregressive Integrated Moving Average (ARIMA) · Artificial Neural Network (ANN)

1 Introduction

Time series forecasting is an active research area that plays important role in planning and decision making in several practical applications [1]. The main task of this research area is to improve the prediction accuracy. For decades, autoregressive integrated moving average (ARIMA) and artificial neural network (ANN) are wildly used for time series prediction. The ARIMA is popular due to capability in dealing with various types of data such as stationary and

© Springer International Publishing AG 2017
V. Torra et al. (Eds.): MDAI 2017, LNAI 10571, pp. 159–169, 2017.
DOI: 10.1007/978-3-319-67422-3_14

non-stationary data. However, the linear relationship between historical and predicted time series is pre-assumed. Such assumption is very difficult to be completely satisfied in practical situations.

On the other hand, the ANN can predict the future time series without any prior assumption. Nevertheless, there is no forecasting model that can be the best for all time series. For instance, the ARIMA works properly for linear time series but for nonlinear time series, the ANN can model nonlinear time series while the ARIMA is not appropriate. Therefore, using the single model is insufficient for dealing with the time series of real-world applications which normally contain linear and nonlinear relationship [2].

Wavelet transform is traditionally used for decomposing signal into low frequency (approximation) and high frequency (detail) components. In time series forecasting, the wavelet transform is adapted to decompose time series. Prediction performances of both ARIMA and ANN have been improved through the wavelet transform in several applications: electrical price [3,4]; short term load [5]; monthly river discharge [6–8]; groundwater level [18]; rainfall and runoff [9,10]; hourly flood forecasting [11]. Furthermore, even though, the wavelet transform has been combined with ARIMA and ANN models [12], however, this approach assumed that the approximation contains only nonlinear component. Such assumption is not practical since the DWT is not a method to transform time series into linear or nonlinear components.

This study proposes a forecasting model capturing both linear and nonlinear components of the detail and the approximation instead of original time series. Firstly, the discrete wavelet transform (DWT) is used to decompose the time series. Then, the hybrid model of ARIMA and ANN are constructed for the approximation and the detail to extract their linear and nonlinear components. Eventually, the final prediction is the combination of the linear and nonlinear components.

The rest of this paper is organized as follows. In Sect. 2, the ARIMA, the ANN and the DWT are introduced. In Sect. 3, the proposed model is described. The experiments and the results are presented in Sect. 4. Finally, Sect. 5 gives the conclusions.

2   Preliminaries

2.1   Autoregressive Integrated Moving Average (ARIMA)

The autoregressive integrated moving average (ARIMA) is a time series forecasting model widely used for several decades because it is capable to handle both stationary and nonstationary data [1]. However, the relationship between historical and forecasted time series is pre-assumed to be linear. The ARIMA has three components which are autoregressive (AR), integration (I) and moving average (MA). In case that the input time series is nonstationary, the integration (I) which is differencing of time series must be performed before further analysis. The mathematical model of the ARIMA can be expressed as:

$$\phi_p(B)(1 - B)^d y_t = c + \theta_q(B)a_t \tag{1}$$

where y_t and a_t denote the time series and random error at period t respectively, c denotes the constant, $\phi_p(B) = 1 - \sum_{i=1}^{p} \phi_i B^i$, $\theta_q(B) = 1 - \sum_{j=1}^{q} \theta_j B^j$, B is the backward shift operator defined as $B^i y_t = y_{t-i}$, ϕ_i and θ_j are autoregressive and moving average parameters respectively, p and q denote the orders of the autoregressive (AR) and the moving average (MA) respectively, and d denotes the degree of differencing.

2.2   Artificial Neural Network (ANN)

The artificial neural network (ANN) is a nonlinear mathematical model mimicking biological neurons [13]. The ANN is popular in time series forecasting due to its flexibility in fitting the relationship between historical and forecasted time series without prior assumption. The structure of the ANN has three kinds of layer such as input, hidden and output layers. In each layer, there are neurons (nodes). Generally, the numbers of layers and nodes are decided based on architect insight in the problem. However, a feed-forward neural network with single hidden layer has been proved that it can be an universal approximator for any continuous function [14]. The feed-forward neural network [15] can be mathematically expressed as:

$$y_t = f\left(b_h + \sum_{h=1}^{R} w_h g\left(b_{i,h} + \sum_{i=1}^{Q} w_{i,h} p_i \right) \right) \tag{2}$$

where y_t denotes the time series at period t, $b_{i,h}$ and b_h denote the biases, f and g denote the transfer functions which are generally the linear and the nonlinear functions respectively, $w_{i,h}$ and w_h denote the connection weights between the layers, Q and R denote the numbers of the input and the hidden nodes respectively.

In this study, the feed-forward neural network with single hidden layer and Levenberg-Marquardt algorithm with Bayesian regularization training algorithm [16] are involved in the experiments.

2.3   Discrete Wavelet Transform (DWT)

The wavelet transform is a technique analyzing both time and frequency of signals simultaneously [17]. This technique decomposes an input signal into two parts: low frequency information (approximation) and high frequency (detail) by using low and high frequency pass filters. For multiple decomposition level, the approximation of the previous decomposition level is the input of higher decomposition level. Actually, there are two main types of the wavelet transform: continuous and discrete wavelet transforms. However, in practical, the time series are discrete and suitable to be decomposed by the discrete wavelet transform (DWT) as:

$$y_t = A_J(t) + \sum_{j=1}^{J} D_j(t)$$
$$= \sum_{k=1}^{K} c_{J,k}\phi_{J,k}(t) + \sum_{j=1}^{J}\sum_{k=1}^{K} d_{j,k}\psi_{j,k}(t)$$

(3)

where y_t denotes the time series at period t; $A_J(t)$ denotes the approximation of the highest decomposition level (J); $D_j(t)$ denotes the detail of decomposition level j; $c_{j,k}$ and $d_{j,k}$ denote the coefficient of the approximation and detail respectively, at decomposition level j and period k; $\phi_{j,k}(t)$ and $\psi_{j,k}(t)$ denote low (approximation) and high (detail) pass filters respectively, at decomposition level j and period k; K denotes the total number of time series J denotes the total level of decomposition.

3 Proposed Forecasting Model

The main idea of the proposed model is using the unique strength of the ARIMA and the ANN in capturing linear and nonlinear components from time series while not assuming the approximation and the detail from the DWT as either linear or nonlinear. The proposed model consists of thee main steps: time series decomposition, extracting linear and nonlinear components, and final forecasting (Fig. 1).

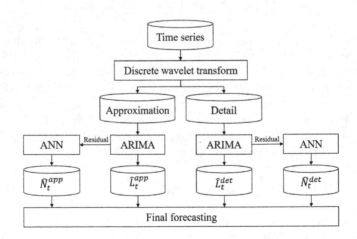

Fig. 1. The proposed forecasting model

In the first step, the DWT with Daubechies wavelet basis function decomposes actual time series (y_t) into the approximation (y_t^{app}), which reveals the trend, and the detail (y_t^{det}) which implied the noise involving seasonality, white noise, etc.

Rather than applying a forecasting model directly to the time series containing both trend and noise, further analysis from the approximation and the detail can give the better prediction result because the forecasting model does not have to deals with the trend and the noise simultaneously.

In the second step, the Zhang's hybrid model of ARIMA and ANN [2] is applied to both the approximation and the detail. This step contributes the new approach that does not make either linear or nonlinear assumption, which have been normally made in several researches, to the characteristic of the approximation and the detail.

Normally, the Zhang's model predicts the future value at period t ($\hat{y}_t$) from a combination of linear ($\hat{L}_t$) and nonlinear ($\hat{N}_t$) components as:

$$\hat{y}_t = \hat{L}_t + \hat{N}_t \tag{4}$$

The linear component ($\hat{L}_t$) can be obtained from the result of applying the ARIMA to the actual time series (y_t). Then the ARIMA residual (e_t) is computed as:

$$e_t = y_t - \hat{L}_t \tag{5}$$

For the nonlinear component ($\hat{N}_t$), it is the result of the ANN using the lagged values of the ARIMA residual (e_t) as the inputs as:

$$\hat{N}_t = f(e_{t-1}, e_{t-2}, \ldots, e_{t-n}) \tag{6}$$

where f is a nonlinear function defined by the ANN, n is total lagged periods.

In case of the proposed model, different Zhang's models are dedicatedly constructed for the approximation and the detail as:

$$\hat{y}_t^{app} = \hat{L}_t^{app} + \hat{N}_t^{app} \tag{7}$$

$$\hat{y}_t^{det} = \hat{L}_t^{det} + \hat{N}_t^{det} \tag{8}$$

where $\hat{y}_t^{app}$ and $\hat{y}_t^{det}$ are the predicted approximation and detail respectively, at period t; $\hat{L}_t^{app}$ and $\hat{L}_t^{det}$ are linear components of the approximation and the detail respectively, at period t; $\hat{N}_t^{app}$ and $\hat{N}_t^{det}$ are nonlinear components of the approximation and the detail respectively, at period t.

The linear components ($\hat{L}_t^{app}$ and $\hat{L}_t^{det}$) are the results of applying the ARIMA to y_t^{app} and y_t^{det} respectively. Then, the ARIMA residuals of the approximation (e_t^{app}) and the detail (e_t^{det}) can be computed as:

$$e_t^{app} = y_t^{app} - \hat{L}_t^{app} \tag{9}$$

$$e_t^{det} = y_t^{det} - \hat{L}_t^{det} \tag{10}$$

For the nonlinear components ($\hat{N}_t^{app}$ and $\hat{N}_t^{det}$), they are obtained from the ANN as:

$$\hat{N}_t^{app} = f^{app}(e_{t-1}^{app}, e_{t-2}^{app}, \ldots, e_{t-n_1}^{app}) \tag{11}$$

$$\hat{N}_t^{det} = f^{det}(e_{t-1}^{det}, e_{t-2}^{det}, \ldots, e_{t-n_2}^{det}) \tag{12}$$

where f^{app} and f^{det} are functions fitted by the ANN, n_1 and n_2 are total lagged periods which are determined by trial and error in the experiments.

At this step, we obtain two linear ($\hat{L}_t^{app}$ and $\hat{L}_t^{det}$) and two nonlinear components ($\hat{N}_t^{app}$ and $\hat{N}_t^{det}$). Finally, the final forecasting can be done by aggregation of the linear and nonlinear components as:

$$\begin{aligned} \hat{y}_t &= \hat{y}_t^{app} + \hat{y}_t^{det} \\ &= \hat{L}_t^{app} + \hat{N}_t^{app} + \hat{L}_t^{det} + \hat{N}_t^{det} \end{aligned} \tag{13}$$

In summary, instead of applying the Zhang's model directly to the time series, the DWT is used for filtering the time series into the approximation (trend) and the detail (noise). Then, without preassuming linear or nonlinear characteristic of the approximation and the detail, the Zhang's model is applied to both of them. In fact, until now, there is no theoretical prove about linear and nonlinear characteristic of the approximation and the detail. Since the specific Zhang's models have been constructed for capturing the trend and the noise separately, they would have more potential to give the better prediction results because the different Zhang's models focus on only either trend or noise (not both of them at the same time). Finally, the final forecasting is done by additive combination of both linear and nonlinear components of the approximation the detail because the relationship between the approximation and the detail is additive as well.

4   Experiments and Results

In order to evaluate the prediction capability of the proposed model, the proposed model is applied to three well-known time series (Table 1): Wolf's sunspot (Fig. 2), Canadian lynx (Fig. 3) and British pound/US dollar exchange rate (Fig. 4). The prediction performance measures involved in this study consist of three measures: mean square error (MSE), mean absolute error (MAE) and mean absolute percentage error (MAPE). The performance of the proposed model is compared to the ARIMA, the ANN and the Zhang's hybrid model.

For the sunspot time series, there are totally 288 annual records (1700–1987). The first 221 records (1700–1920) are used as training set. The remaining 67 records (1921–1987) are test set. Firstly, the sunspot time series is passed through the DWT to generate the approximation and the detail. Secondly, the ARIMA is fitted to both the approximation and the detail. ARIMA(0, 0, 6) and ARIMA(0, 0, 3) are the most fitted models. Thirdly, the residuals of the two most fitted ARIMAs are computed and analyzed by the ANN. The best ANNs for the residuals of the approximation and the detail are ANN(10-10-1) and ANN(2-2-1).

After the final forecasting, the performance measures of short term (35 years) and long term (67 years) horizontal predictions are evaluated (Table 2). From the comparison, the proposed model gives the lowest error in all three measures. In the short term prediction, the MSE, MAE and MAPE are 121.12, 6.16 and 19.45% respectively. For the long term prediction, the MSE, MAE and MAPE are 199.24, 7.77 and 22.56% which are higher than the short term prediction because the long term prediction includes the highest peak at period 37 (see Fig. 5a) that causes shift up in variability of the time series. However, the proposed model is still the best model because the measures of the other models are increased as well. Therefore, for the sunspot, the proposed model is the best model in both short and long term prediction.

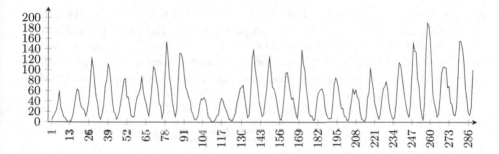

Fig. 2. Sunspot time series (1700–1987)

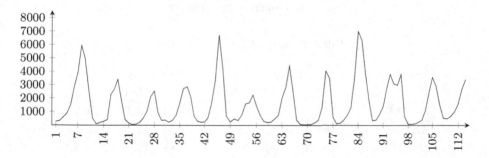

Fig. 3. Canadian lynx time series (1821–1934)

The Canadian lynx time series contains 114 observations (1821–1934). The size of the training and the test set are 100 observations (1821–1920) and 14 observations (1921–1934) respectively. After obtaining the approximation and the detail from the DWT, their best fitted ARIMAs are determined as ARIMA(0, 0, 5) and ARIMA(2, 0, 0) respectively. The most suitable ANNs for predicting the residuals of these two ARIMAs are ANN(3-8-1) and ANN(4-4-1) respectively. The performance comparison is shown in Table 3. The proposed model

Fig. 4. Exchange rate time series (1821–1934)

has the best performance in MSE, MAE and MAPE which are 0.0154, 0.1014 and 3.4575% respectively. With the proposed model, the MSE is significantly improved. Based on the mathematical formular of the MSE, it is more sensitive to a big error. Thus, lower MSE has more chance to promise lower maximum of error. In this case, the proposed model has the lowest maximum error which is at period 1 (see Fig. 5b).

Table 1. Detail of time series and experiment

Time series	Size (total, training, test)
Sunspot (1700–1987)	(288, 221, 67)
Canadian lynx (1821–1934)	(114, 100, 14)
Exchange rate (1980–1993)	(731, 679, 52)

Table 2. Sunspot forecasting result

Model	35 year ahead			67 year ahead		
	MSE	MAE	MAPE	MSE	MAE	MAPE
ARIMA	197.87	10.52	29.17%	323.48	13.25	32.86%
ANN	164.08	9.51	31.76%	413.90	14.19	33.34%
Zhang's model	156.76	9.63	30.22%	300.88	12.74	32.08%
Proposed model	**121.12**	**6.16**	**19.45%**	**199.24**	**7.77**	**22.56%**

Table 3. Lynx forecasting result

Model	MSE	MAE	MAPE
ARIMA	0.0229	0.1120	3.7062%
ANN	0.0201	0.1165	4.0156%
Zhang's model	0.0247	0.1083	3.5504%
Proposed model	**0.0154**	**0.1014**	**3.4575%**

Table 4. Exchange rate forecasting result

Model	1 month			6 months			12 months		
	MSE	MAE	MAPE	MSE	MAE	MAPE	MSE	MAE	MAPE
ARIMA	36.6514	0.01508	3.6628%	35.5387	0.0155	3.8883%	29.0517	0.01375	3.4286%
ANN	33.7797	0.0147	3.5767%	37.6072	0.0157	3.9298%	30.4533	0.01404	3.5065%
Zhang's model	36.1826	0.0151	3.6605%	35.6239	0.0156	3.8994%	28.8004	0.01379	3.4335%
Proposed model	**31.0259**	**0.0102**	**2.5541%**	**10.9023**	**0.0073**	**1.8240%**	**8.1812**	**0.00601**	**1.4959%**

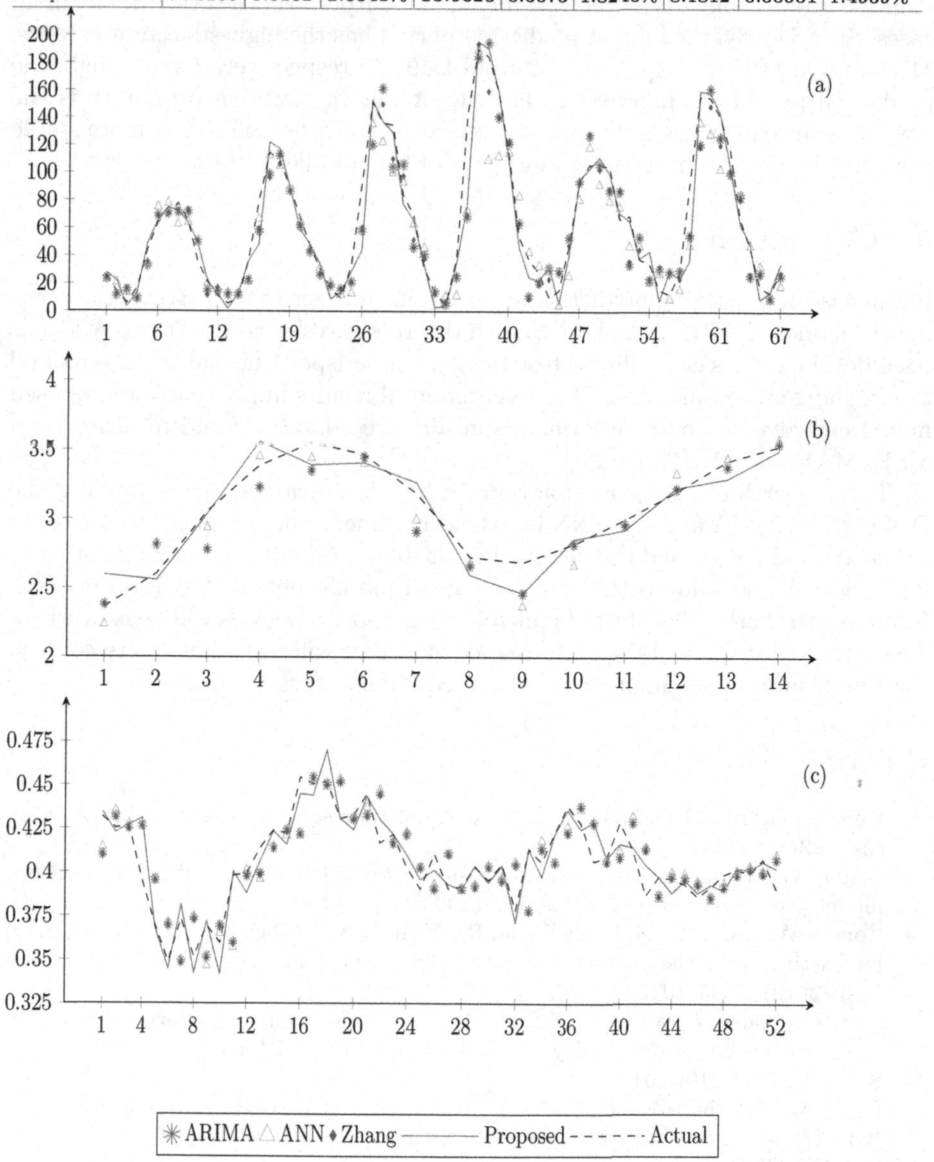

Fig. 5. Forecasted values: (a) Sunspot, (b) Canadian lynx, (c) Exchange rate

In case of the exchange rate, the data set includes 731 weekly exchange rates (1980–1993). The data is partitioned into 679 and 52 observations as training and test set. The ARIMAs fitted to the approximation and the detail are ARIMA(0, 1, 0) and ARIMA(0, 0, 3) respectively. The suitable ANNs of the ARIMAs residuals are ANN(2-3-1) and ANN(7-9-1). In evaluation of the prediction performance, three horizontal predictions are considered: 1 month, 6 months and 12 months. From Table 4, the proposed model can outperform the other model in all cases. Surprisingly, the longest prediction period has the highest accuracy; MSE, MAE and MAPE are 8.1812, 0.00601 and 1.4959% respectively. Eventhough, the prediction period (12 months) is the longest but the average variability is the lowest because it includes the end of the year (see Fig. 5c) which has more stable exchange rates than both the beginning and the middle of the year.

5   Conclusions

In purpose of improving prediction accuracy in time series forecasting, the novel hybrid model of ARIMA and ANN with discrete wavelet transform (DWT) has been developed. Its capability is tested with the sunspot, the Canadian lynx and the exchange rate time series. The experimental results imply that the proposed model can give the best performance in all three data sets and measures (i.e. MSE, MAE and MAPE).

The better forecasting accuracy indicates the advantage of combining the DWT, the ARIMA and the ANN in extracting linear and nonlinear components of the approximation and the detail without linear or nonlinear assumption. The limitation of this study is that the decomposition level of the DWT is only one. In the future works, the effect of multiple decomposition levels will be concerned. The statistical testing of the performance measures will be conducted to confirm the significance of the improvement in prediction accuracy.

References

1. De Gooijer, J.G., Hyndman, R.J.: 25 years of time series forecasting. Int. J. Forecast. **22**(3), 443–473 (2006)
2. Zhang, G.P.: Time series forecasting using a hybrid ARIMA and neural network model. Neurocomputing **50**, 159–175 (2003)
3. Conejo, A.J., Plazas, M.A., Espinola, R., Molina, A.B.: Day-ahead electricity price forecasting using the wavelet transform and ARIMA models. IEEE Trans. Power Syst. **20**(2), 1035–1042 (2005)
4. Tan, Z., Zhang, J., Wang, J., Xu, J.: Day-ahead electricity price forecasting using wavelet transform combined with ARIMA and GARCH models. Appl. Energy **87**(11), 3606–3610 (2010)
5. Fard, A.K., Akbari-Zadeh, M.R.: A hybrid method based on wavelet, ANN and ARIMA model for short-term load forecasting. J. Exp. Theor. Artif. Intell. **26**(2), 167–182 (2014)
6. Zhou, H.C., Peng, Y., Liang, G.H.: The research of monthly discharge predictor-corrector model based on wavelet decomposition. Water Resour. Manag. **22**(2), 217–227 (2008)

7. Wei, S., Zuo, D., Song, J.: Improving prediction accuracy of river discharge time series using a wavelet-NAR artificial neural network. J. Hydroinformatics **14**(4), 974–991 (2012)
8. Adamowski, J., Sun, K.: Development of a coupled wavelet transform and neural network method for flow forecasting of non-Perennial rivers in semi-arid watersheds. J. Hydrol. **390**(1), 85–91 (2010)
9. Nourani, V., Komasi, M., Mano, A.: A multivariate ANN-wavelet approach for rainfall-runoff modeling. Water Resour. Manag. **23**(14), 2877–2894 (2009)
10. Partal, T., Kişi, O.: Wavelet and neuro-fuzzy conjunction model for precipitation forecasting. J. Hydrol. **342**(1), 199–212 (2007)
11. Tiwari, M.K., Chatterjee, C.: Development of an accurate and reliable hourly flood forecasting model using WaveletBootstrapANN (WBANN) hybrid approach. J. Hydrol. **394**(3), 458–470 (2010)
12. Khandelwal, I., Adhikari, R., Verma, G.: Time series forecasting using hybrid ARIMA and ANN models based on DWT decomposition. Procedia Comput. Sci. **48**, 173–179 (2015)
13. Zhang, G., Patuwo, B.E., Hu, M.Y.: Forecasting with artificial neural networks: the state of the art. Int. J. Forecast. **14**(1), 35–62 (1998)
14. Hecht-Nielsen, R.: Theory of the backpropagation neural network. In: International Joint Conference on Neural Networks, IJCNN 1989, pp. 593–605 (1989)
15. Dayhoff, J.A.: Neural Network Architectures: An Introduction. MIT Press, Cambridge (1995)
16. MacKay, D.J.: A practical Bayesian framework for backpropagation networks. Neural Comput. **4**(3), 448–472 (1992)
17. Fliege, N.J.: Multirate Digital Signal Processing: Multirate Systems, Filter Banks, Wavelets. Wiley, Chichester (1994)
18. Adamowski, J., Chan, H.F.: A wavelet neural network conjunction model for groundwater level forecasting. J. Hydrol. **407**(1), 28–40 (2011)

A Novel Hybrid Model for Activity Recognition

Hela Sfar[(⊠)], Amel Bouzeghoub, Nathan Ramoly, and Jérôme Boudy

CNRS Paris Saclay, Telecom SudParis, SAMOVAR, Paris, France
{hela.sfar,amel.bouzeghoub,nathan.ramoly,
jerome.boudy}@telecom-sudparis.eu

Abstract. Activity recognition focuses on inferring current user activities by leveraging sensory data available. Nowadays, combining data driven with knowledge based methods has show an increasing interest. However, uncertainty of sensor data has not been tackled in previous hybrid models. To address this issue, in this paper we propose a new hybrid model to cope with the uncertain nature of sensors data. We fully implement the system and evaluate it using a large real-world dataset. Experimental results prove the high performance level of the proposal in terms of recognition rates.

Keywords: Smart home · Machine learning · Ontology · Activity recognition

1 Introduction

Context awareness is a hot research field in various domains since it may help to ensure end-user well being and improve quality of life. A trending context aware applications are the smart environments, in particular smart home (SH). SH have emerged as an achievable technology that can support peoples, such as the elderly, for independent living. To provide assistance for individual inhabitants of an SH, activity recognition is needed to identify the task that the individual is currently undertaking based on the received sensor data. However, in real world deployment, sensors data are not always well-aimed and precise [7] as devices may encounter multiple breakdown including hardware failure, energy depletion, communication problems, etc. Therefore, a context-aware system should handle the missing or imprecise sensors data in order to make the right decisions. Previously, approaches tackling sensors' uncertainty to recognize activity can be into three categories: *data driven*, *knowledge based*, and *hybrid*. Data driven methods [7,9,10] apply different supervised and unsupervised machine learning techniques to classify sensors data into activities based on provided training datasets. Support Vector Machine (SVM) [9], and Conditional Random Fields (CRF) [10] are examples of well-known classifiers. Although this group of methods is suited to handle uncertainty and to deal with a broad range of sensors, it also needs a large amount of training data to set up a model and estimate its parameters. On the other hand, knowledge based methods relies on

© Springer International Publishing AG 2017
V. Torra et al. (Eds.): MDAI 2017, LNAI 10571, pp. 170–182, 2017.
DOI: 10.1007/978-3-319-67422-3_15

logic rules, ontologies and reasoning engine to infer proper activities from current sensors input. Despite their powerful semantic representation of real world data and their reasoning capabilities, their use is restricted to a limited number of sensors. Given the limitations of both data driven and knowledge based approaches, combining them is a promising research direction as it was stated in analysis done in [7,11]. Intuitively, a hybrid approach takes the "best of both worlds" by using a combination of methods. Such an approach is able to provide a formal, semantic and extensible model with the capability of dealing with uncertainty of sensors data and reasoning rules [7]. Therefore, proposing hybrid models has been the motivation of recent works including [2,12]. However, none of the aforementioned approaches of hybrid models address the problem of sensors data uncertainty. This a clear drawback in our context. To overcome this limitation, this paper proposes a new hybrid model combining data driven and knowledge based methods for activity recognition and uncertainty management. The main contributions of the paper are as follows:

1. Introduction of a new hybrid model for activity recognition in smart home that combines the ontology and semantic reasoning with data driven technique. The novel model handles uncertain sensors data and it exploits these uncertainty values to compute the produced activities uncertainty values.
2. Improvement of an existing method for feature extraction [5] in order to deal with more time distant actions and their uncertainty values.
3. Fully implemented system for activity recognition with a high level of efficiency.

The rest of this paper is organized as follows: Sect. 2 discusses related work about hybrid models for activity recognition. Section 3 presents the proposed hybrid model. Section 4 reports experimental results. Finally, Sect. 5 concludes the paper.

2 Related Work

The combination of data driven and knowledge based methods for activity recognition has been a recent topic of interest. Therefore, few hybrid activity recognition systems have been proposed in the literature.

COSAR [3,13] is a context-aware mobile application that combines machine learning techniques and an ontology. As a first step, the machine learning method is triggered in order to predict the most probable activities based on a provided training data. Then, an ontological reasoner is applied to refine the results by selecting the set of possible activities performed by a user based on his/her location acquired by a localization server. Despite the fact that the sensor data are supposed to be certain, COSAR deals with the uncertainty of the transformation of the localization from a physical format to a symbolic one. Another hybrid model that combines a machine learning technique, an ontology, and a log-linear system has been proposed in [12]. The aim of this approach is to recognize a multilevel activity structure that holds 4 levels: atomic gesture (Level 4), manipulative gesture (Level 3), simple activity (Level 2), and complex activity (Level 1). The atomic gestures are recognized through the application of a machine learning technique. Moreover, using a probabilistic ontology defined by the log-linear, and standard ontological reasoning tasks, the manipulative gestures, simple activities, and complex activities are inferred. Each level is deduced based on a time window that contains elements from the previous level. Even though the work in [12] is similar to the previous one regarding the absence of sensor data uncertainty's handling, the inference of the 4 levels activities is based on a probabilistic reasoning that represents a sort of uncertainty.

FallRisk [8] is another pervasive system that combines data driven and knowledge based methods. Its main objective is to detect a fall of an elderly person living independently in a smart home. FallRisk is a platform that integrates several systems that use machine learning methods for fall detection. It filters the results of these systems thanks to the use of an ontology that stores the contextual information about the elderly person. The main advantage of this system is that it is extensible to integrate several fall detection systems. Moreover, the contextual information of the elderly is taken into account. However, this work does not consider any kind of uncertainty.

FABER [14] is a pervasive system used to detect abnormal behavior of a patient. Firstly, it deduces events and actions from the acquired sensor data. This is done based on simple ontological inference methods. Then, these events and actions are sent to a Markov Logic Network (MLN) as a machine learning method to analyze the event logs and infer the start/end time of activities. The inferred activity boundaries are communicated together with actions and events to the knowledge based inference engine. This engine evaluates the rules modeling abnormal behaviors and detected abnormal behaviors are communicated to the hospital center for further analysis by the doctors. Nevertheless, similarly to previous works this system does not handle uncertainty of sensor data.

SmartFABER [2] system is an extension and an improvement of FABER [14]. These two frameworks share the same aims. Regarding SmartFABER [2], instead of communicating the inferred events and actions to MLN classifier, the system sends them to a module that is in charge of building vectors of features based on the received events. Then, these features are communicated to a machine learning module for the classification of activities. Next, a proposed algorithm called SmartAggregation is applied to infer current activity instances. For deducing an activity's instance from a sequence of events classified to an activity, the algorithm verifies whether each event satisfies a set of conditions. These conditions are defined by a human expert after a deep analysis of the semantic of activity. If all events satisfy all conditions, then an activity instance could be inferred. This work is proved to outperform FABER [14]. However, it suffers from two main drawbacks: (1) There is no uncertainty handling for sensor data; (2) The performance of the SmartAggregation algorithm depends heavily on the defined conditions. It can suffer from time consuming if there is a huge number of conditions that need semantic verification.

Combining the Hidden Markov Model (HMM) with the ontology has been also addressed in [17] to recognize the user activity within a wireless smart home. In the proposed prototype system, the smart devices are described by semantic models. When the user needs assistance, smart gateway can provide appropriate services according to the inference results of the HMM algorithm. However, the sensor data uncertainty issue has not been tackled in this method. In [18] a recent proposal for activity recognition that mixes the use of ontology and the statistical learning approach Markov Logic Network (MLN). The proposed system utilizes the model theoretic semantic property of description logic, to convert the represented ontology activity model to its corresponding first order rules. MLN is constructed by learning weighted first order rules that enable probabilistic reasoning within a knowledge representation framework. However, the sensor data uncertainty handling is not discussed enough in the paper.

3   Model Architecture

The model architecture contains, as depicted in Fig. 1, two main components: Ontological modeling & Semantic reasoning, and Data Driven layer. The former is in charge of the ontological representation of sensors data and their uncertainty values. Afterwards, it infers events and actions from this representation and compute their uncertainty values. The latter, receives the actions together with their uncertainty values. It starts by gathering the input into features and computes the features weights (Features extraction). Then, it classifies actions and features into activities (Activities classification). In the following, clear explanation of each component is given.

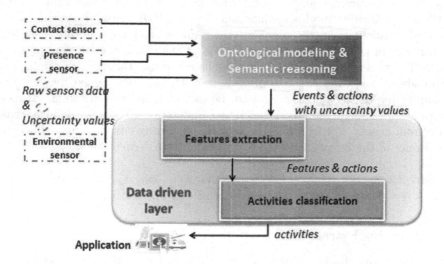

Fig. 1. Architecture overview

3.1   Ontological Modeling and Semantic Reasoning

The main aim of this module is to infer events and actions and compute their uncertainty values based on the coming sensors data with their uncertainty values. In this work we consider three levels of inference. The lower one contains the events which are the simple atomic gesture done by the user. The second level is the actions that are about more complicated gestures of the user. The third level is inferred by the data driven layer that contains the set of activities. An activity is the conclusion of a set of actions that result the undertaking task of the user. The ontology is used in our model to represent the sensors data together with their uncertainty values. This ontological modeling allows sensors data to be formally conceptualized. This conceptualization, as the ontology is defined for, serves to provide a semantic model from these data. Moreover, it becomes possible to infer new information based on these modeled data. To do so, for events and actions inference, simple logic rules are applied. The uncertainty values of the produced events and actions are computed thanks to the application of the possibility logic [4]. Regarding events, they are deduced from sensors data. For a better insight we show how we can deduce the occurrence of an event with label *SitOnChair*:

$\forall$ t_1, $t_2 \in \{Time\}$, p $\in \{Person\}$, n_1, $n_2 \in \{Uncertainty\}$
$\langle p; hasLocomotion; SitOn\rangle \land \langle SitOn; hasUncertainty; n_1\rangle \land \langle SitOn; hasTime;$
   $t_1\rangle \land$
$\langle Chair; hasState; Used\rangle \land \langle Chair; hasUser, p\rangle \land \langle Used; hasUncertainty; n_2\rangle \land$
$\langle Used; hasTime; t_2\rangle \longrightarrow ev(SitOnChair, max(t_1, t_2), min(n_1, n_2))$

As we can see, the produced event is in the form *ev(e, t, c)*. Where *e* is its label, *t* is its time occurrence and *c* its uncertainty value. In the above example, the left side of the rule contains conjunction between RDF triplet representing the received sensor data in the ontology, their uncertainty values, and their time occurrence. In the right side, the conclusion of the rule is the creation of a new event. The uncertainty value of this event is computed through the application of possibility logic and equal to $min(n_1, n_2)$. This is because the rule contains a conjunction between clauses. Due to the lack of space we refer to [4] for more information about possibility logic. Regarding actions, they are deduced basically from events. For example through the following rule we can deduce the occurrence of the actions with label *SitOnChairAtKitechenTable*:

$ev(SitOnChair, t_1, n_1) \land ev(PresenceAtKitchenTable, t_2, n_2) \land (t_1 \geq t_2) \land$
$((t_1 - t_2) \leq 5s) \longrightarrow (SitOnChairAtKitchenTable, t_2, min(n_1, n_2))$

3.2   Features Extraction

The second contribution of this paper, as we stated in the introduction, is a proposal of new measure to compute the weight of the extracted features. This measure is to overcome the issues in a previous measure proposed in [5]. More clearly, for each received action $ac(a_i, t_i, c_i)$, this module is in charge of building a feature vector representing the sequence S of the recent actions in a time window's size $\mathbf{n}$: $S = \langle ac(a_{i-n+1}, t_{i-n+1}, c_{i-n+1}),..., ac(a_{i-1}, t_{i-1}, c_{i-1}), ac(a_i, t_i, c_i)\rangle$. In this work, we improve the technique proposed in [5]. We have chosen this method since it is proved to be effective in features extraction based on sensors events or actions streams instead of streams of sensors data compared to traditional features extraction techniques [7]. This technique builds a vector of feature for each events sequence S. The produced feature vector of an events sequence S_i holds essentially: (i) The label of the feature, K_i; (ii) The list of events under S_i; (ii) A weight value fine-tunes the contribution of each event in S_i, so that recent events participate more than the older ones. This weight value is computed as follows:

$$F_{k_i}(S_i) = \sum_{ev(e_j,t_j)\in S_i} exp(-\chi(t_i - t_j)) \times f_{k_i}(ev(e_j,t_j)) \tag{1}$$

where the factor χ determines the time-based decay rate of the events in S_i; $t_i - t_j$ is expressed in seconds and $f_{k_i}(ev(e_j, t_j))$ is the time-independent participation of $ev(e_j, t_j)$ in the computation of the F_{k_i} value. The other way around if $ev(e_j,t_j)$ participates in the execution of k_i then $f_{k_i}(ev(e_j,t_j)) = 1$ and else 0. For a better illustration, Fig. 2 shows the curve's shape of the function $exp(-\chi(t_i - t_j))$ with $\chi = 0.5$ and , $t_i - t_j \in [0..10]$. As we can see in the figure, when $t_i - t_j \succ 4s$, $exp(-\chi(t_i - t_j)) \approx 0$. Consequently, $ev(e_j, t_j)$ with $t_j \succ 4s$ does not participate in the computation of F_{k_i}. In our understanding, this may be true when the approximate duration of the feature is short. In contrast, when it is long, this hypothesis is not always valid. Let us take the example of the feature "stove usage" in [2]. The duration of the execution of this feature, for a particular recipe, may be equal to a number of hours. We suppose that the vector of this feature contains the following two events "openStove" and "closeStove". Since the duration of the feature "stove usage" is in terms of hours, the value of $(t_{closeStove} - t_{openStove})$ may be equal to 1 or 2 h. Accordingly, by applying Eq. 1 the event "openStove" does not participate in the feature "stove usage" weight calculus. Therefore, incorrect value of weight may be obtained. Accordingly, the execution of the event "openStove" must have a high impact in the execution of the feature. Intuitively, the stove can not be used if it is not opened. Moreover, this technique assumes that the only factor that may have impact on the computation of the feature weight is the time distance between events. However, when

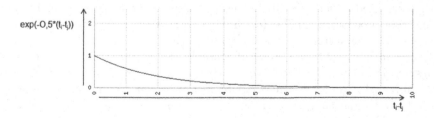

Fig. 2. The graphic of $exp(-\chi(t_i - t_j))$ with $\chi = 0.5$

information about actions and events uncertainty values is provided, this information should participate in the computation of the feature's weight. Therefore we made the following assumptions:

A1. The uncertainty values must have an impact on the calculus of the feature weight: the higher uncertainty value of event or action is, the more the weight of the feature increases.

A2. The uncertainty of an event or action value can be decreased, but it cannot be increased.

To overcome the problem mentioned above, we propose a new version of Eq. 1 that meets the assumptions A1 and A2. We formally define the notion of Short Time Feature (STF) and Long Term Feature (LTF) as follows:

Definition 1. *Short Time Features (STF) is a set of features that holds only features having duration less or equal to ten minutes. Let $Dur(f)$ be the duration of feature f: $f \in STF \Leftrightarrow Dur(f) \leq 600s$.*

Definition 2. *Long Term Features (LTF) is a set of features that holds only features having duration more than ten minute. Let $Dur(f)$ be the duration of feature f: $f \in LTF \Leftrightarrow Dur(f) \succ 600s$.*

Firstly, it is important to note that this value of ten minutes is chosen intuitively and it is variable according to the experiments. Then, to compute the weight of the feature, we propose the following Eq. 2.

$$F_{k_i}(S_i) = \sum_{ac(a_j,t_j,c_j) \in S_i} c_j \times Fact_{\chi,\delta t_{ij}} \times f_{k_i}(ac(a_j,t_j,c_j)) \tag{2}$$

$$Fact_{\chi,\delta t_{ij}} = \begin{cases} exp(-\chi * \delta^h t_{ij}) \; If \; (k_i \in LTF) \\ \frac{1}{\chi * \delta^s t_{ij}} \quad If((\chi * \delta^s t_{ij} \succ 1) \; \& \\ \qquad\qquad (k_i \in STF)) \\ 1 \qquad\qquad Otherwise \end{cases}$$

Foremost, it is worth mentioning that in our work the features vectors are built from actions instead of events in contrast to [5]. We note that $c_j \times Fact_{\chi,\delta t_{ij}}$, in Eq. 2, is a sort of uncertainty where $c_j \times Fact_{\chi,\delta t_{ij}} \leq c_j$ (Assumption A2). $\delta t_{ij} = t_i - t_j$. $\delta^h t_{ij}$ means that δt_{ij} is expressed in hours and $\delta^s t_{ij}$ means that δt_{ij} is expressed in seconds. To fix the problem of the time delay in Eq. 1, we distinguish three cases: (1) The feature is a LTF ($k_i \in LTF$). Then, to compute $Fact_{\chi,\delta t_{ij}}$ the same function $(exp(-\chi * \delta t_{ij}))$ in Eq. 1 is used. However, δt_{ij} is expressed in hours ($\delta^h t_{ij}$) instead of seconds. Accordingly, $\delta^h t_{ij}$ will have low values and then the curve will have less decreasing shape. Therefore, the actions that happened earlier could participate in the weight computation in contrast to Eq. 1; (2) The feature is a STF ($k_i \in STF$) and $\chi * \delta t_{ij} \geq 1$. Then, $Fact_{\chi,\delta t_{ij}} = \frac{1}{\chi * \delta^s t_{ij}}$. The quotient function is chosen to compute $Fact_{\chi,\delta t_{ij}}$ since it has a lesser decreasing shape than the exponentiation function. It is obvious as it can

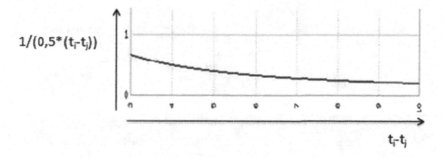

Fig. 3. The graphic of $\frac{1}{\chi * \delta^s t_{ij}}$ with $\chi = 0.5$

be seen in Fig. 3 which shows the shape of the curve for the function $\frac{1}{\chi * \delta^s t_{ij}}$ with $\chi = 0.5$ and $\delta^s t_{ij} \in [3..10]$; (3) If $(\chi * \delta^s t_{ij}) \prec 1$, then $\frac{1}{\chi * \delta^s t_{ij}} \succ 1$ and accordingly $c_j \times Fact_{\chi, \delta t_{ij}} \succ c_i$ that does not correspond to Assumption A2. Therefore, $Fact_{\chi, \delta^s t_{ij}} = 1$. The reason behind this is: since t_i and t_j are very close, $ac(a_j, t_j, c_j)$ must have the highest impact on the weight computation, i.e. the value 1 in [0..1] is chosen. The computed weights serve, subsequently, as their uncertainty values. Afterwards, the actions and the features together with their uncertainty values are sent to `Dempster-Shafer theory for activity classification` module.

3.3 Activities Classification

In order to classify activities, we propose a new model for the application of the Dempster Shafer theory (DS) [1] to classify actions and features into activities. DS has proven to provide decent results in comparison to other machine learning techniques such as J48 Decision Tree [6]. Usually the Directed Acyclic Graph (DAG) is used to represent the named source evidences in DS, their hypothesis, the mass functions, the activities, and to support the distribution and the fusion of evidences. In the DAG, evidence sources represent the root nodes at the base of the diagram. Evidence source readings are mapped to one or more hypothesis. Each one in turn will be mapped to one or more activities. In this work, as depicted in Fig. 4, the DAG is used where each evidence source is a sequence of actions Si which are in the feature vector. Then the named hypothesis in DS are the features in our work. The mass function value is computed as the number of the feature's occurrence while the activity execution. These mass function values are discounted by being multiplied with the feature weight value F_{k_i} (Eq. 2) after normalization. This product reflects an uncertainty value about the production of the feature (its weight value) and about the classification of the features into activities (mass value). The final value of uncertainty is propagated to activities thanks to the Dempster's rule of combination. We refer to this paper [1] for more details about this rule. This final value forms the uncertainty value of the matching activity. Please note that Dempster Shafer could be replaced by any

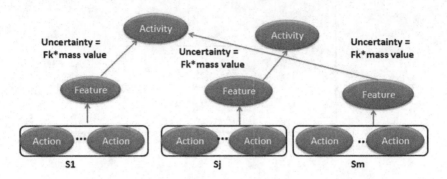

Fig. 4. DAG for activity recognition applied in our model

classification method that support uncertainty handling, such as Support Vector Machine [9] (SVM), see Sect. 4.

4   Experimental Evaluation

For this experimentation, we evaluate the proposed system. In this section, we first describe the dataset used for the experiments. Then, the experiments and the achieved results are presented.

4.1   DataSet

We used real-life data collected in highly rich smart environments. The dataset[1] was obtained as a part of the EU research project "Activity and Context Recognition with Opportunistic Sensor Configuration"[2]. The dataset holds (from our model perspective) sensors data (level 4), events (level 3), actions (level 2), and activities (level 1) that have been done by three persons S10, S11, and S12 with three different routines each (ADL1-3). In order to test our system, this dataset does not contain information about sensors data uncertainty values (level 4). Therefore, we have randomly annotated the level 4 in the dataset with high uncertainty values in [0.8...1]. Moreover, we have injected a set erroneous sensors data in the dataset annotated with low uncertainty values in [0...0.4]. In the experimentation, we have tested the performance of the system with different number of injected erroneous sensors data.

4.2   Implementation and Experimental Setup

We have implemented the proposed system using JAVA. Regarding data driven layer, for this dataset, we have considered the set of features depicted in Table 1

---

[1] The dataset can be downloaded from this link: http://webmind.dico.unimi.it/care/annotations.zip.

[2] http://www.opportunity-project.eu.

to be treated in the *Time & Uncertainty based features extraction* module. Since the *Time & Uncertainty based features extraction* module requires the preliminary step in which the value of parameter n, corresponding to the time window duration of the temporal sequence of actions, is experimentally chosen.

Table 1. List of considered features. STF: Short Time Feature, LTF: Long Term Feature

Feature name	STF/LTF	Duration (s)
PrepareCoffee	STF	600
Drink	STF	120
GatherCutlery	STF	600
GatherFood	STF	600
Eat	STF	1200
PutAwayFood	STF	600
PutAwayCutl	STF	600
Dishwhasher	STF	300
Resting	STF	600

Therefore, we have tested the method with different values of n. Furthermore, for the *Activity classification* module, the experimentation is not limited only to the use of Demspter Shafer Theory – we also have the Support Vector Machine (SVM) [9] for activity classification. The results obtained from this module are compared with those obtained in [12].

4.3   Evaluation and Results

As described in the previous sections, the *Activity classification* module outputs the predicted activity class. It is worth mentioning that the system has no False Negative result (FN = 0): it outputs always at least one activity. Figure 5 depicts the average precision measure over the three routines for all subjects by varying the value of n in [60s...300s] with one erroneous sensor data for five correct ones (e.g., 1/5 erroneous sensors data). We have chosen the interval [60s...300s] for varying the value of n since all the features for this dataset are STF (as presented in Table 1). Therefore, the time window n must not exceed the maximum duration of an STF (e.g., 600s see Subsect. 4.1) You can find more detailed result online[3]. As it is clearly shown in the Fig. 5, the Dempster Shafer (DS) with n = 180 s reaches 91% of precision recognition rate. For time windows shorter or longer than 180, DS tends to become less efficient: DS is efficient where time window is properly proportioned to the activities: when the time window is too

---

[3] http://nara.wp.tem-tsp.eu/what-is-my-work-about/cognition-situation-reaction-wip/agacy/.

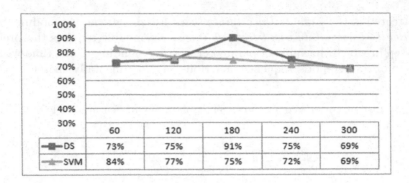

Fig. 5. Average recognition precision for all subject over the three routines with different values of the time window n

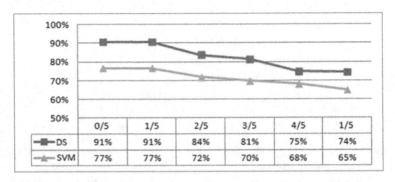

Fig. 6. Average recognition precision for all subject over the three routines for varying frequency of uncertain event. The value 1/5 means there is one uncertain sensor data for five correct ones. 1 means there is one uncertain sensor data for one correct one

big, there may be conflict between activities, while when they are too small, DS does not have enough data to be efficient. On the other hand, SVM gives better results for short time window ($n \le 120\,s$), but with the increase of n value, the accuracy of classification gets worse. This maybe explained by the fact that SVM is not as dependent on features weight as DS. In general, DS provides better results than SVM.

Figure 6 shows the average precision measure over the three routines for all subjects by varying the proportion of the introduced uncertain sensors data compared to the correct sensors data with n = 180 s. As it is clearly shown in the figure the DS [1] theory is more efficient than the SVM [9]: The precision values of DS [1] are in range [0,74...0,91], however that of SVM are in range [0,65...0,77]. Both methods have a decreasing precision values when the number of uncertain sensors data in the dataset increases. This is an expected result since the methods will have less certain data to make the right decision. However, the system even with a dataset whose half contains uncertain sensors data it can predict the activity with a good precision level (74%).

5 Conclusion

In this paper, we proposed a new hybrid model for activity recognition and sensors data uncertainty handling. The main novelty of our proposal is the combination of knowledge based with data driven methods. Thus, several modules contribute to compute the uncertainty value of the expected output. Unlike the related work, the proposed system supports the inherent uncertain nature of sensors data and exploits it to compute the uncertainty value of each module's output. Furthermore, our experiments show the proposed system to be efficient and precise even with uncertain sensor data. Comparing the model results with other previous hybrid model, proposing algorithm for inferring activities instances, and proposing an ontological model for uncertainty integrating are the extensions of the proposed method [16].

Acknowledgements. This work has been partially supported by the project COCAPS (https://agora.bourges.univ-orleans.fr/COCAPS/) funded by Single Interministrial Fund N20 (FUI N20).

References

1. Lotfi, A.Z.: A simple view of the Dempster-Shafer theory of evidence, its implication for the rule of combination. AI Mag. **7**, 85–90 (1986). ACM
2. Riboni, D., Bettini, C., Civitares, G., Janjua, Z.H.: SmartFABER: recognizing fine-grained abnormal behaviors for early detection of mild cognitive impairment. AI Med. **67**, 47–57 (2016). Elseiver
3. Riboni, D., Bettini, C.: COSAR: hybrid reasoning for context-aware activity recognition. Pers. Ubiquit. Comput. **3**, 271–289 (2011). Springer
4. Dubois, D., Lang, J., Prade, H.: Automated reasoning using possibilistic logic: semantics, belief revision, and variable certainty weights. IEEE Trans. Knowl. Data Eng. **6**, 64–71 (1994). IEEE
5. Krishnan, N.C., Cook, D.J.: Activity recognition on streaming sensor data. Pervasive Mob. Comput. **10**, 138–154 (2014). Elseiver
6. Sebbak, D., Benhammadi, F., Chibani, A., Amirat, Y., Mokhtari, A.: Dempster-Shafer theory-based human activity recognition in smart home environments. Ann. Telecommun. **69**, 171–184 (2014). Springer
7. Ye, J., Dobson, S., McKeever, M.: Situation identification techniques in pervasive computing: a review. Pervasive Mob. Comput. **9**, 36–66 (2012). Elseiver
8. De Backere, F., Ongenae, F., Van den Abeele, F., Nelis, J., Bonte, P., Clement, E., Philpott, M., Hoebeke, J., Verstichel, S., Ackaert, A., De Turck, F.: Towards a social and context-aware multi-sensor fall detection and risk assessment platform. Comput. Biol. Med. **64**, 307–320 (2015). Elseiver
9. Hearst, M.A., Dumais, S.T., Osuna, E., Platt, J., Scholkopf, B.: Support vector machines. IEEE Intell. Syst. Appl. **13**, 18–28 (1998)
10. Buettner, M., Prasad, R., Philipose, M., Wetherall, D.: Recognizing daily activities with RFID-based sensors. In: ACM International Joint Conference on Pervasive and Ubiquitous Computing (2009)
11. Ziaeefard, M., Bergevin, R.: Semantic human activity recognition: a literature review. Int. J. Pattern Recogn. **48**, 2329–2345 (2015). Elseiver

12. Helaoui, R., Riboni, D., Stuckenschmidt, H.: A probabilistic ontological framework for the recognition of multilevel human activities. In: ACM International Joint Conference on Pervasive and Ubiquitous Computing (2013)

13. Riboni, D., Bettini, C.: Context-aware activity recognition through a combination of ontological and statistical reasoning. In: International Conference on Ubiquitous and Intelligence Computing (2009)

14. Riboni, D., Bettini, C., Civitarese, G., Janjua, Z.H., Helaoui, R.: Fine-grained recognition of abnormal behaviors for early detection of mild cognitive impairment. In: IEEE International Conference on Pervasive Computing and Communication (2015)

15. Anil, J., Karthik, N., Arun, R.: Score normalization in multimodal biometric systems. Int. J. Pattern Recogn. **38**, 2270–2285 (2005). Elseiver

16. Sfar, H., Bouzeghoub, A., Ramoly, N., Boudy, J.: AGACY monitoring: a hybrid model for activity recognition and uncertainty handling. In: Extended Semantic Web Conference (2017)

17. Guo, K., Li, Y., Lu, Y., Sun, X., Wang, S.Y., Cao, R.H.: An activity recognition-assistance algorithm based on hybrid semantic model in smart home. Int. J. Distrib. Sensor Netw. **12** (2016). Sage Journal

18. Gayathri, K.S., Easwarakumarb, K.S., Elias, S.: Probabilistic ontology based activity recognition in smart homes using Markov logic network. Knowl. Based Syst. **121**, 173–184 (2017). Elseiver

Constructing Document Vectors Using Kernel Density Estimates

Michael Mayo[✉] and Sean Goltz

Department of Computer Science, University of Waikato, Hamilton, New Zealand
{michael.mayo,sean.goltz}@waikato.ac.nz

Abstract. Document vector embeddings are numeric fixed length representations of text documents that can be used for machine learning and text mining purposes. We describe in this paper a new technique for generating document vectors. Our novel idea builds on the recently popular notion of neural word vector embeddings and combines this concept with the statistics of kernel density estimation. We show that robust document vectors can be produced using our new algorithm, and perform an experiment involving several challenging text classification datasets to demonstrate its effectiveness.

1 Introduction

Machine learning approaches to text-based classification problems usually rely on suitable preprocessing methods to convert the text of a document into a numeric representation suitable for further processing. Typically document lengths vary, and preprocessing methods often account for this by producing fixed-length representations.

A classic approach taken for document preprocessing is the Bag-of-Words (BOW) method [20]. Essentially, this approach considers all possible words in the corpus and represents a document as a binary vector where each 1 denotes the presence of a particular word, and each 0 denotes its absence. Commonly occurring words that may lack informative value such as "is" and "a" (stop words) may be removed, thus reducing the dimensionality of the vectors. The method can be extended by replacing the binary values with frequencies instead, or some metric computed from the frequencies such as the term frequency–inverse document frequency, which may increase overall performance (according to machine learning metrics such as accuracy).

Clear disadvantages of this approach, however, are threefold. Firstly, the order of the words is ignored, which may result in loss of the information contained in word orderings (e.g. consider the difference in meaning of "house boat" vs. "boat house"). Secondly, the dimensionality of the document vectors depends on the size of the corpus – the greater the number of documents in the corpus, the larger the quantity of words, in general. And thirdly, no prior information about the words are assumed beyond their occurrence statistics in the corpus being used for training.

© Springer International Publishing AG 2017
V. Torra et al. (Eds.): MDAI 2017, LNAI 10571, pp. 183–194, 2017.
DOI: 10.1007/978-3-319-67422-3_16

More recently, neural word embeddings [1,10,13,14] have been proposed as a means of overcoming primarily the third (but also the second) of these disadvantages. The basis for neural word embeddings is the so-called distributional hypothesis [7] stating that words with similar meanings frequently occur in similar contexts. Thus, unsupervised learning can be performed using unlabelled text (e.g. news articles or Wikipedia text) prior to supervised learning in order to determine clues about word semantics based on context. Traditional BOW approaches typically ignore this information.

The basic idea behind neural word embeddings is as follows: if a neural network with a single hidden layer can be used to either predict the context of a word $w_1, \ldots w_{i-1}, w_{i+1}, \ldots w_n$ given the word w_i (the Skip-Gram model [13,14]) or alternatively predict a word w_i given its context $w_1, \ldots w_{i-1}, w_{i+1}, \ldots w_n$ (the Continuous Bag-of-Words approach [13]), then the activation values of the hidden layer of the network can serve as a representation of the current word. This is a word vector.

Producing document vectors can be straightforward when neural word embeddings are available. The simplest approach is averaging the word vectors in the document. Since the size of the hidden layer is fixed prior to training, then the size of the corresponding word vectors is also fixed. Averaging of word vectors in a document is a powerful, simple and practical method for computing document vectors. To illustrate, [9] performed one comparison between word vector averaging and a much more complex recursive neural network approach for the purposes of text classification. It was shown that the averaging approach is competitive with the significantly neural network approach, and therefore we employ word vector averaging as the baseline algorithm in this research.

The main focus of this paper is a new method for constructing document vectors. Our idea is based on the entirely novel notion of modelling a document's words using Kernel Density Estimation (KDE, [19]). Essentially, the word vectors of a document can be modelled using a KDE, and then the KDE can be sampled at regular intervals throughout the word embedding space to produce a fixed-length representation of a document. This is an alternative and computationally simpler approach than some of the other ideas in the literature.

We show in a comprehensive set of text classification experiments that this approach leads to statistically significant improvements in performance compared to a baseline method that produces document vectors via simple word vector averaging.

2   Background

In this section, the notion of word and document vectors are reviewed in more detail and then the basis of KDE is outlined.

2.1   Word and Document Vectors

Devising algorithms for constructing useful new word embeddings is an ongoing research area. It is currently an open question as to which methods result in the most effective word embeddings. [12] performed a rigorous comparison between several competing modern approaches and concluded that parameter tuning contributes significantly to the differences in performances, and this may have significant impact on experimental results: therefore there is currently no clear single "best" approach.

An early neural word embedding approach is [13] who proposed both the CBOW and the skip-gram models for learning word vectors. The skip-gram model was shown to be the most effective of the two and is frequently the focus of later studies such as those reported in [14].

Since then, several refinements to word vector learning algorithms have been proposed. Amongst them, the `fastText` algorithm and its corresponding open source implementation [1,10] significantly reduces the training time of the neural embedding model by several orders of magnitude. The system achieves this by learning character n-gram vectors instead of word vectors. Word vectors are then constructed indirectly by summing the n-gram vectors that form a word [1].

The `fastText` approach has been shown empirically to outperform most existing approaches to text classification on most problems while at the same time significantly speeding up both training time by an order of magnitude [10].

2.2   Kernel Density Estimation

KDE is an approach from statistics for estimating the underlying (hidden) probability distribution that a set of data samples are drawn from [3,19]. In statistics, low-dimensional KDEs are useful for visualisation purposes, while in machine learning, KDEs find useful application in clustering (e.g. the DENCLUE algorithms [8]).

In essence, KDE starts with the choice of a kernel function K. There are many possible choices for K, and in this paper we set K to the oft-used Gaussian distribution with zero mean and unit variance, i.e.

$$K(y) = \frac{1}{\sqrt{2\pi}} e^{-\frac{y^2}{2}} \tag{1}$$

Given K and a set of data samples $x_1, x_2, \ldots x_n$, a one-dimensional KDE is defined as

$$f(x|h) = \frac{1}{nh} \sum_{i=1}^{n} K\left(\frac{x - x_i}{h}\right) \tag{2}$$

where h is the bandwidth parameter governing the smoothness of the KDE. This parameter is chosen to ensure that the f is neither too smooth nor that it overfits the samples [3].

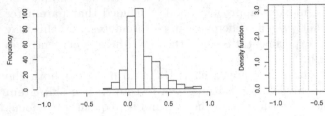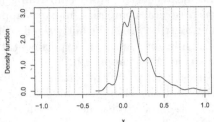

Fig. 1. Construction of a frequency histogram (a) and its corresponding KDE (b) from 252 data points. Vertical lines in b indicate a fixed number of uniformly-spaced samples of the KDE as discussed in Sect. 3.

Figure 1a shows a frequency histogram for a set of 252 samples drawn from an unknown underlying probability distribution. The reconstruction of the underlying distribution using KDE is shown in Fig. 1b.

3 Kernel Density Estimation for Defining Document Vectors

In this section, we outline a novel approach to generating document vectors based on KDE, which is the main contribution of this paper. Figure 2 summarises the approach graphically.

We first of all assume that each document d in our corpus can be represented by a bag of words, $\{w_1, w_2, \ldots w_{|d|}\}$. In turn, a function $f(w) \rightarrow v$ which converts words into vectors of fixed constant dimensionality n is also assumed. The function f can be thought of as encapsulating a neural embedding method such as a skip gram model trained using `fastText`. The dimensionality will typically be quite high for these vectors, e.g. $n = 100$ is a reasonable value. The bag of words for d can then be converted into a corresponding bag of vectors $\{v_1, v_2, \ldots v_{|d|}\}$.

A document can therefore be represented as a matrix with $|d|$ rows, one per word, and n columns. Now consider a single column from this matrix. This column, with $|d|$ values, can be thought of as a random sample of values drawn from some underlying probability distribution. Figure 1 illustrates this process with the variable x corresponding to the first column of a matrix constructed from a document with 252 words.

Given that there are n columns in this matrix and since neural word embeddings typically set n to be quite large in order for them to be effective, this will result in a large number of one-dimensional KDEs being constructed.

Once the n KDEs are constructed, the next step of our approach is to sample at fixed and regular intervals the KDEs in order to produce a "resampled" set of data. Let us assume that there are e such resamples (or estimates) per dimension, with $e \ll |d|$. Two significant outcomes of this step are: (i) data reduction, i.e. we are converting the $|d| \times n$ matrix representation of the document into a $e \times n$ matrix, and (ii) length equalisation, i.e. if e is a constant across all documents then the resulting matrix will be the same size for all documents regardless of the varying number of words between documents.

This estimation process is illustrated by Fig. 1b. In the figure, a KDE constructed from 252 samples is estimated at 21 equally-spaced points between -1 and 1, as indicated by the vertical dotted lines. Thus the number of data points required to model the document along one dimension has been reduced from 252 to 21. If this same process is repeated for each of the n dimensions of the word vectors, then if $n = 100$, the document representation is reduced from $252 \times 100 = 252,000$ original data points (which may not be uniformly spaced in embedding space) to 2,100 estimates that are uniformly spaced. We hope that this uniform resampling process will produce a more concise and accurate "picture" of the document's contents than the original word vectors do.

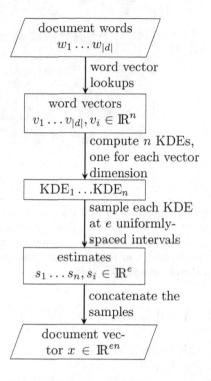

Once the reduced matrix has been computed, it is possible to visualise it in order to obtain some insight into the distribution of its elements. Figure 3 illustrates this for the same document that was used to generate the example in Fig. 1b. The same number of estimates ($e = 21$) per dimension is used as before. As can be observed, the distributions across each of the $n = 100$

Fig. 2. The KDE-based approach to a computing a vector for a single document.

dimensions varies quite a bit, with values ranging from 0 to a maximum of nearly 3.5. The matrix appears to be semi-sparse, with values near the edges of each sequence (i.e. close to -1 or 1) frequently zero or very close to zero. Furthermore, each dimension clearly has a differently-shaped KDE.

The primary advantage of the reduced matrix is that it is a compressed, fixed-size representation of the document that ideally captures all essential features of the document. The next step therefore is to flatten the matrix into a 1D vector of length $e \times n$, which can then be input to standard machine learning algorithms.

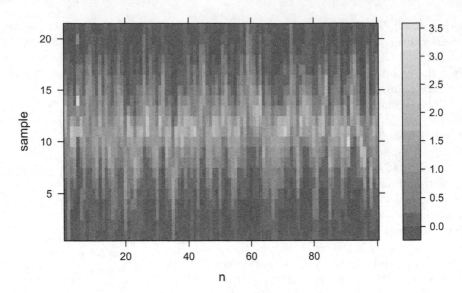

Fig. 3. Visualisation of a document vector representation of a 252-word article from the BBC Sports dataset. The n axis indicates the dimension of the word vector and the *sample* axis indicates the sample number (along that axis, sample 0 corresponds to position -1 and sample 21 corresponds position 1; the step size is 0.1). The intensity indicates the sample value.

4   Evaluation

In this section, we describe an extensive evaluation of the document vector representation method outlined in the previous section. We use word vector averaging as a baseline document vector representation and consider six challenging document classification datasets along with three different machine learning algorithms.

4.1   Datasets

Six text classification datasets were selected for our evaluation in this paper. Table 1 gives some basic statistics about each of the datasets, in particular the number of classes, number of documents, and the average document length in words. Some comments on each of the datasets are also briefly given.

BBC News. The first dataset listed in Table 1 is the BBC News dataset, first contributed to the research community by [6]. The dataset consists of over two thousand news articles from the BBC and is divided into five classes: business, entertainment, politics, sport and technology. The problem to solve is news topic identification.

Table 1. Description of the text datasets used for evaluations.

Dataset	Num. documents	Num. classes	Average words/Doc
BBC news	2225	5	384
BBC sports	737	5	341
Compliance	80	2	110
Political ideology	4335	3	42
Movie reviews	2000	2	746
Deceptive reviews	800	2	119

BBC Sports. The BBC Sports dataset (also courtesy of [6]) is similar to the news dataset just mentioned in that it involves news topic identification, but the number of examples is smaller. Furthermore, this dataset focusses on sports news articles only, dividing articles into five classes namely athletics, cricket, football, rugby and tennis. Both the news and sports datasets from the BBC are approximately evenly balanced in terms of numbers of examples per class.

Canadian Compliance Law. The Canadian Compliance Laws dataset is a small dataset created by the authors for a previous study [5]. The documents in this dataset are fragments or clauses of Canadian federal and provincial legislation. The negative examples are clauses from laws and regulations that do not contain penalties. The positive examples are clauses taken from regulations that do contain penalties. Both clause types contain dollar amounts, and the examples in this dataset are evenly balanced between the classes. Due to the infrequency of positive examples in this domain, this dataset is relatively small compared to the other datasets.

Political Ideology. The political ideology dataset is a challenging dataset [9] consisting of a large number of very small fragments of text, with the task being to label each fragment as either conservative, liberal or neutral. The dataset is severely imbalanced with 1701 fragments labelled conservative, 2025 fragments labelled liberal, but only 600 neutral fragments. In original paper in which this dataset was introduced [9], significant preprocessing steps were taken improve the accuracy prior to experiments being performed. Specifically, these were: (i) the dataset was subsampled to make the classes more evenly balanced, and (ii) phrase level annotations were added to the examples to increase the amount of explicit information used for training. In order to keep our experiment simple and reproducible, we do not perform the same preprocessing steps, but instead run our experiment on the entire unaltered dataset.

Movie Reviews. The movie reviews dataset version 2.0 [17] consists of 2,000 reviews of movies taken from popular online blogs and magazines. The reviews are evenly divided between positive reviews (i.e. the reviewer liked the movies) and negative reviews, so this is a sentiment classification task. The document length is much longer on average than the other datasets.

Deceptive Hotel Reviews. The opinion spam dataset version 1.4 [15,16] poses a similarly challenging problem to the movie review dataset. The problem that the dataset encapsulates is to detect untruthful reviews of hotels, otherwise known as "opinion spam". The dataset itself is divided into four categories, each consisting of 400 reviews. The categories are: positive truthful reviews of a hotel, positive deceptive reviews, negative truthful reviews, and finally negative deceptive reviews. A hierarchical approach to classification could be taken for this problem, in that one could build multiple classifiers (e.g. a top level classifier for positive vs. negative sentiment followed by two lower level classifiers for distinguishing truthful vs. deceptive reviews). However, as with the political ideology dataset, we kept our approach here simple since these are initial experiments and instead divide the examples into two evenly balanced classes: truthful and deceptive.

4.2   Statistical and Machine Learning Algorithm Implementations and Parameter Settings

We use several publicly available open source software implementations to perform our experiments. For learning the word vectors, we make use of the fastText system [1,10][1] trained on a dump of the English-language version of Wikipedia[2]. The following text preprocessing steps were taken to prepare the text: firstly, only the visible portions of the Wikipedia dump were preserved (i.e. XML tags were removed); and secondly, numbers were converted into their word equivalents (i.e. "200" was converted to "two zero zero").

The settings used for our fastText experiments are all standard defaults (i.e. skipgram model, word vector dimensionality $n = 100$, learning rate of 0.025 etc.) except for the number of epochs which is set to four to improve the quality of the word vectors.

For computing the one-dimensional KDEs, we use the R package ks [3], version 2.23, again with all default settings. The range and intervals for taking estimates from each KDE are the same as shown in Fig. 1b, i.e. 21 estimates per dimension are taken between -1 and 1 inclusive in steps of 0.1.

Finally, the machine learning algorithms we utilise are implemented in WEKA 3.8.0 [4]. Particularly, we use an implementation of Support Vector Machines (SVMs) [11] in conjunction with two different kernels: a simple linear kernel and a radial basis function kernel for producing a non-linear classifier. All settings for the SVM and the kernels are defaults. Besides SVMs, we also tested the well known Random Forests [2] ensemble classifier, but increased the

---

[1] Version from December 2016.
[2] http://mattmahoney.net/dc/textdata.html as at December 2016.

Table 2. Experiment results (mean testing accuracy and standard deviation)for the topic modelling datasets.Symbols ∘, • indicate statistically significant improvement or degradation with 95% certainty.

Classifier	BBC news		BBC sports	
	WV averaging	KDE approach	WV averaging	KDE approach
SVM (linear)	97.01(1.03)	**97.48(0.96)**	93.46(2.72)	**97.61(1.58)** ∘
SVM (radial basis)	95.16(1.38)	97.09(1.14) ∘	49.83(3.58)	94.47(2.37) ∘
Random forest	95.92(1.27)	95.23(1.36)	84.76(3.71)	80.66(3.71) •

Table 3. Experiment results (mean testing accuracy and standard deviation) for the compliance and political ideology datasets. Symbols ∘, • indicate statistically significant improvement or degradation with 95% certainty.

Classifier	Compliance		Political ideology	
	WV averaging	KDE approach	WV averaging	KDE approach
SVM (linear)	**96.50(5.91)**	90.50(10.07)	54.42(2.11)	50.26(2.09) •
SVM (radial basis)	90.75(10.60)	76.75(12.05) •	48.29(0.81)	**54.81(1.96)** ∘
Random forest	90.25(10.89)	93.50(8.04)	52.57(1.73)	50.93(1.50) •

Table 4. Experiment results (mean testing accuracy and standard deviation) for the opinion-based datasets. Symbols ∘, • indicate statistically significant improvement or degradation with 95% certainty.

Classifier	Movie reviews		Deceptive reviews	
	WV averaging	KDE approach	WV averaging	KDE approach
SVM (linear)	74.14(3.23)	77.91(2.70) ∘	81.03(2.90)	84.87(2.32) ∘
SVM (radial basis)	69.23(3.10)	**79.93(2.70)** ∘	76.59(3.08)	**85.26(2.48)** ∘
Random forest	70.39(3.04)	72.58(3.23)	77.94(3.03)	79.84(3.05)

ensemble size from its WEKA default value to 10,000. At the same time we fixed the number of randomly-selected attributes per tree to five.

4.3   Results

The experiments we conducted were stratified ten-fold cross validation experiments repeated ten times. Exactly 100 train/test runs were therefore performed for each algorithm/dataset/document vector type combination. For each dataset and algorithm, the resulting accuracies were averaged and the standard deviations computed.

The results are given in Tables 2, 3 and 4 (we have split the results into multiple tables to enhance the readability) and are laid out so that the accuracies using the word vector averaging approach (the baseline) can be compared directly to our new KDE-based approach.

We also performed standard statistical t-tests at 95% confidence to determine if the document vector type makes any significant difference to the average classification accuracy. The results of this test are noted in the tables.

Examining the results, we can observe first of all that the KDE-based approach to defining document vectors improves the accuracy of the classification algorithm compared to the using the baseline document vectors in each case except for the Compliance dataset.

On the BBC News and Political Ideology datasets, the improvement is small. In the case of the news dataset, this appears to be because word vector averaging already achieves a high accuracy; in the case of the ideology dataset, the problem appears to be simply extremely challenging. For that particular dataset, all methods achieved a performance not much greater than would be expected by a simple method that simply classifies all fragments as "conservative" (which would yield approximately 46% accuracy). The difficulty of classification using this dataset is most likely a result of the extremely short average document length (42 words per example compared to 384 words for the news dataset) and the typical nature of political language, which may be more subtle or ambiguous then other language.

However, performance on the BBC Sports dataset as a consequence of applying our new document vector algorithm results in a much more drastic improvement, going from a best result of approximately 93% accuracy to a new best result of approximately 97% – quite a significant performance gain.

Similar improvements are observed with the Movie Reviews and the Deceptive Hotel Reviews datasets. The best performing movie review sentiment detector using a KDE-based document vector representation has nearly 80% accuracy compared to word vector averaging which achieves at most approximately 74% accuracy. Similarly, the gain in performance accuracy for the problem of detecting fraudulent hotel reviews is in the order of 3% as a direct consequence of our new approach.

After considering these results, we revisited the compliance dataset. The best performing classifier was a linear SVM in conjunction with word vector averaging, which achieved a mean best accuracy of 96.5%. We hypothesise that our KDE-based approach did not improved performance in this one case because of overfitting: the dimensionality of the KDE-based word vectors is 2,100 but the number of examples in the dataset is 80, significantly less than the other datasets. Machine learning algorithms typically find such "wide" datasets challenging.

We therefore re-ran our experiments on this dataset only, but modified the algorithms so that prior to the classifier being constructed, we first of all selected the best 100 feature values according to the information gain of each feature with respect to the class. This was performed using the training data only for each run. This selection process resulted in 2,000 feature values from the document vectors being discarded, and significantly reduced the dimensionality of the KDE-based document vectors for the compliance dataset. As a consequence, the KDE-based approach in conjunction with a linear SVM improved its accuracy to $98.00 \pm 4.61\%$, a reasonable (although not statistically significant) improvement.

5   Conclusion

To conclude, we have provided in this paper a new technique for document vector construction based on modelling a document's "bag of word vectors" using KDEs, and evaluated it in the context of document classification using machine learning.

The results show that the in most cases the degree of improvement is positive, but this depends significantly on the problem domain and the machine learning classifier used. For some domains (e.g. BBC News) the improvement is small; for others (e.g. the sentiment domains) the improvement is quite significant. Future work in this area could focus on an analysis of which domain and classifier characteristics lead to improved performance, and why. Future work could also explore in more detail one of the domains, either a domain where the algorithm works well compared to the baseline (e.g. Movie Reviews) to further improve and understand performance, or one where performance is poor (i.e. Political Ideology) in order to better ascertain why.

Another issue is the complexity of this algorithm compared to the baseline. For the datasets used in our experiments, the time complexity of computing the KDEs was minimal because the datasets were small to medium sized. However, it is an open question as to how the complexity would scale to very large problems with millions of documents as our method requires multiple KDEs to be computed for each document. A critical issue here is how the bandwidth parameter h (in Eq. 2) is chosen. In our current implementation, h is determined using a combination of optimisation and cross validation which is an effective method but may not scale well. A more efficient alternative would be to use Silverman's rule of thumb [18] and see how much this impacts performance. Using kernels other than Gaussian may also improve the KDE computation complexity.

Finally, future refinements to the algorithm itself are also possible. For example, we could address the selection of parameters used to construct both the KDEs and the number and location of the estimates along the KDE used as feature values. In the case of the former, we have used default parameter settings in the ks package, and in the case of the latter we have used a fixed range of $[-1, 1]$ in steps of 0.1 for all datasets. Optimising these values for specific problems may be helpful, and also allow us to address the issue of document vector size, which is 2,100 for our KDE-based approach compared to 100 for the word vector averaging approach.

References

1. Bojanowski, P., Grave, E., Joulin, A., Mikolov, T.: Enriching word vectors with subword information. arXiv preprint arxiv:1607.04606 (2016)
2. Breiman, L.: Random forests. Mach. Learn. **45**(1), 5–32 (2001)

3. Duong, T.: **ks**: kernel density estimation and kernel discriminant analysis for multivariate data in R. J. Stat. Soft. **21**(7), 1–16 (2007)
4. Frank, E., Hall, M., Witten, I.: The WEKA workbench. In: Data Mining: Practical Machine Learning Tools and Techniques, 4th edn. Morgan Kaufmann (2016)
5. Goltz, S., Mayo, M.: Enhancing regulatory compliance by using artificial intelligence text mining to identify penalty clauses in legislation. In: Proceedings of Workshop on MIning and REasoning with Legal Texts (MIREL 2017) (2017, to appear)
6. Greene, D., Cunningham, P.: Practical solutions to the problem of diagonal dominance in kernel document clustering. In: Proceedings of 23rd International Conference on Machine Learning (ICML), pp. 377–384 (2006)
7. Harris, Z.: Distributional structure. Word **10**(23), 146–162 (1954)
8. Hinneburg, A., Gabriel, H.-H.: DENCLUE 2.0: Fast clustering based on kernel density estimation. In: R. Berthold, M., Shawe-Taylor, J., Lavrač, N. (eds.) IDA 2007. LNCS, vol. 4723, pp. 70–80. Springer, Heidelberg (2007). doi:10.1007/978-3-540-74825-0_7
9. Iyyer, M., Enns, P., Boyd-Graber, J., Resnik, P.: Political ideology detection using recursive neural networks. In: Proceedings of 52nd Annual Meeting of the Association for Computational Linguistics (ACL), pp. 1113–1122 (2014)
10. Joulin, A., Grave, E., Bojanowski, P., Mikolov, T.: Bag of tricks for efficient text classification. arXiv preprint arxiv:1607.01759 (2016)
11. Keerthi, S., Shevade, S., Bhattacharyya, C., Murthy, K.: Improvements to Platt's SMO algorithm for SVM classifier design. Neural Comput. **13**(3), 637–649 (2001)
12. Lau, J., Baldwin, T.: An empirical evaluation of **doc2vec** with practical insights into document embedding generation, Technical report arxiv:1607.05368 (2016)
13. Mikolov, T., Chen, K., Corrado, G., Dean, J.: Efficient estimation of word representations in vector space. Technical report, arXiv preprint arxiv:1301.3781 (2013)
14. Mikolov, T., Sutskever, I., Chen, K., Corrado, G., Dean, J.: Distributed representations of words and phrases and their compositionality. In: Advances in Neural Information Processing Systems 26 (NIPS 2013) (2013)
15. Ott, M., Cardie, C., Hancock, J.: Negative deceptive opinion spam. In: Proceedings of 2013 Conference of the North American Chapter of the Association for Computational Linguistics (2013)
16. Ott, M., Choi, Y., Cardie, C., Hancock., J.: Finding deceptive opinion spam by any stretch of the imagination. In: Proceedings of 49th Annual Meeting of the Association for Computational Linguistics (2011)
17. Pang, B., Lee, L., Vaithyanathan, S.: Thumbs up? Sentiment classification using machine learning techniques. In: Proceedings of Empirical Methods on Natural Language Processing, pp. 79–86 (2002)
18. Silverman, B.: Density Estimation for Statistics and Data Analysis. Chapman & Hall/CRC, London (1986)
19. Simonoff, J.: Smoothing methods in statistics. Springer, New York (1996). doi:10.1007/978-1-4612-4026-6
20. Turney, P., Pantel, P.: From frequency to meaning: vector space models of semantics. J. Artif. Intell. Res. **37**, 141–188 (2010)

On Cell Detection System with User Interaction

Yoshitaka Maeda[1]([✉]), Yasunori Endo[2], and Sho Sanami[1]

[1] Dai Nippon Printing Co., Ltd., Shinjuku, Japan
maeda-y12@mail.dnp.co.jp
[2] Faculty of Engineering, Information and Systems,
University of Tsukuba, Tsukuba, Japan

Abstract. Cell detection is the essential step for various analyses in biological fields. One of the conventional approach is to construct a specialized method for every specified task, but it is not efficient in the meaning of the coding workload. Another approach is based on some machine learning technique, but it is difficult to prepare many training data. To solve these problems, we propose a balanced system by combining image processing and machine learning. The system is universally applicable to any image, because it only consists of basic methods of image processing. The code of the system does not need to be modified, because its behavior is adaptively tuned by machine learning. Users are free from excessive request of training data, because only a few desirable data is specifically requested by the system. The system consists of three units to achieve functionalities for avoiding parameter collision, compensating lack of training data, and reducing complexity of feature space. The effectiveness of the system is evaluated with a typical set of cell images, and the result is sufficient. The proposed system provides a reasonable way of preparing a tuned cell detection method for arbitrary sets of images in this field.

1 Introduction

Analysis of shape or movements of cells is useful for a wide range of fields such as biology and pathology. In many cases, the cell regions are detected by image processing at the beginning of the analysis task. A large amount of image processing techniques have been developed in various fields, e.g. post-processing on photograph and robot vision. Each technique such as filter operations, image normalization, edge extraction is applicable to the cell detection [9,10], if their translucency is carefully considered. However, it is necessary to prepare a different algorithm for each task, because the appearance of the cells varies according to the type of cell, magnification of microscope, state of illumination, and so on. It is inefficient to write algorithms for every task in the fields.

Hence many researches have been done aiming to provide universal or flexible methods for cell detection. For example, the cell region can be obtained by solving some optimization problem based on the mechanism of microscopic imaging [1]. Although this approach is very effective for some types of microscope, the same idea cannot be applied to the most common bright field microscope.

© Springer International Publishing AG 2017
V. Torra et al. (Eds.): MDAI 2017, LNAI 10571, pp. 195–206, 2017.
DOI: 10.1007/978-3-319-67422-3_17

There are also researches to apply deep neural networks (DNN) to detect cell regions [3,4]. DNN-related technologies have been improved their performance remarkably today, but sometimes they require too much on training data to get satisfying results. It is difficult to apply DNN to cell detection except in special situations where large amounts of data can be prepared.

The goal of this paper is to propose a cell detection system which can handle many types of images with reasonable amount of user workload. The desired characteristics of the system are as follows:

- It can be applied to different tasks without any modification on code.
- It does not rely on characteristics which are specific to a certain type of cell.
- It does not require heavy user workload, in terms of quality and quantity.

2 Proposed System

2.1 Overview of the System

We propose a practical system for cell detection based on conventional image processing and machine learning. The overview of the system is shown as Fig. 1. The system consists of three processing units as follows:

Assortment Unit: Input images are classified.
Drafting Unit: Rough cell regions are detected.
Reformation Unit: The detected rough cell regions are improved.

The first unit acts like an algorithm switcher for stability and efficiency of the system. The characteristics of the cell images may change unexpectedly even within a single biological experiment. In addition, the images may contain abnormal images due to device errors. If these images are processed without discrimination, the result of the analysis will be undesirable. In the proposed system, each input image is classified into an appropriate image-class or excluded as an error. Each image-class assigns its own cell detection process. Thus the system can operate stably and efficiently for usual image-sets of cells.

The latter two units perform as a specialized cell-detector for each image class. Machine learning techniques can be useful for this specialization. However, it is difficult to acquire a good criterion with small amount of training data because the feature space of the image is too complicated. Worse, it is difficult to identify the datum which improves the criterion even if the user is positive to add training data. In order to solve these problems, the detection process of proposed system consists of two units; drafting unit and reformation unit. In the drafting unit, rough cell regions are detected. The rough detection is performed based on some classwise associations defined by user inputs. When the association is ambiguous, adequate questions are generated to resolve the ambiguity. This behavior compensates for lack of training data efficiently. In the reformation unit, rough cell regions are improved by supervised learning. The improvements are performed restrictively along the boundary of the regions. This restriction reduces the complexity of the feature space. Therefore, detection processes are specialized efficiently even with minimal user inputs and simple classifiers.

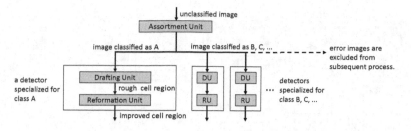

Fig. 1. Overview of the system

2.2   Usage of the System

Before explaining the details of the units, we will describe about the usage of the system. The main usage is to recognize the cells in an image, considering the type of the image. Therefore the system has to be trained about the type before the recognition performed. The overview of the training process and the recognition process are shown as Fig. 2. Each unit has two functions; for training and for recognition.

In the training process (Fig. 2a), multiple images are processed at once with some user interactions. The training process is constructed as follows:

1. In the first unit, the recognition function is trained by all input images.
2. The input images are classified into subsets by the recognition function. The subsequent learning process is repeated for each of them.
3. In the second unit, the recognition function is trained with the each subset. The function is improved with some user interactions as necessary.
4. The rough cell regions are detected in each image by the recognition function. The image and its rough cell regions are coupled as an output.
5. In the third unit, the recognition function is trained with the outputs. The recognition function is updated with some user interactions.

In the recognition process (Fig. 2b), each image is processed one-by-one without any user interaction. The recognition process is just a sequence of the recognition function of each unit.

2.3   Assortment Unit

In this unit, the system classifies the input images into a small number of groups for efficient training and stable operation.

The training process consists of image-normalization, calculation of the image-feature, and formulation of image-classification criterion, as shown in Algorithm 1. The image-feature F_{img} is a vector which is calculated for every normalized image I_{nrm}. Hierarchical clustering is performed on all image-features, and some clusters are extracted by a given threshold. An image-class is defined for each cluster by a center vector and a radius, i.e. the centroid and the standard deviation of the image-features in the cluster. The criterion $\mathcal{L}_{image}$ consists of these image-class definitions.

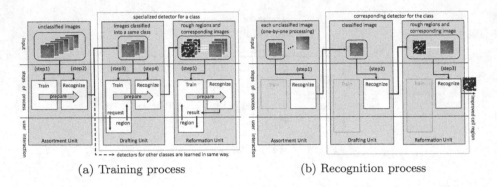

(a) Training process   (b) Recognition process

Fig. 2. Usage of the proposed system

In the recognition process, the input images are discriminated into the appropriate image-class or error, using the criterion $\mathcal{L}_{\text{image}}$ as shown in Algorithm 2. A candidate class c_{in} is identified so that the distance between the image-feature F_{img} and the center vector $\text{Center}[i]$ is minimized. If the minimum distance is less than the reference radius of the criterion, the candidate class c_{min} is accepted. Otherwise, the image is regarded as an error and excluded from subsequent processes.

About Pseudo-Codes in This Paper. The outline of the algorithm is shown using MATLAB-like and Python-like notation. In particular, the binary operator such as .* or ./ denotes the component-wise operations. Variables decorated with @ denotes that they are given parameters. Any variable in script capitals like $\mathcal{L}$ or $\mathcal{FG}$ denotes that the variable's scope is global. Since the name of function itself shows the main concept of operation, the details of trivial function is not provided. However, the implementation for our experiment would be briefly described in Sect. 3.2. The same rules apply to other algorithms shown in this paper.

2.4 Drafting Unit

In this unit, the system extracts rough cell regions by a detection criterion. The criterion is based on a pixel-wise classification using a pixel-feature. The pixel-feature is a vector which can be calculated from the small area around the pixel. The training process of the drafting unit consists of two phases, as shown in Algorithm 3.

The first phase is to create multiple criteria for pixel-classification. A set of pixel-features are collected so that they approximately cover the entire pixel variations of all input images. Next, k-means clustering is applied to the set using various k as cluster number. A pixel-class is defined from each cluster by using some measure, such as centroid and standard deviation of the pixel-features in the cluster. Each pixel-classification criterion M_k is composed of these definitions of pixel-classes.

Algorithm 1. Training Process in Assortment Unit

```
void Learn_Assortment ( List of Image images )
    features = List of ImageFeature
    foreach Image I_org ∈ images
        I_nrm = Normalize ( I_org )
        F_img = CalculateImageFeature ( I_nrm )
        features.Add ( F_img )
    L_image = ImageClassifier
    clusters = HierarchicalClustering ( features ).ExtractClusters (@threshold)
    foreach Index i of clusters
        L_image.Center[i] = Mean ( v ∈ clusters[i] )
        L_image.Radius[i] = Std ( v ∈ clusters[i] ) * @scale
end
```

Algorithm 2. Recognition Process in Assortment Unit

```
ImageClassIndex Recognition_Assortment ( Image I_org )
    I_nrm = Normalize ( I_org )
    F_img = CalculateImageFeature ( I_nrm )
    c_min = arg min ‖L_image.Center[i] − F_img‖
              i
    if ‖L_image.Center[c_min] − F_img‖ < L_image.Radius[c_min]
        then return c_min
        else return ERROR_CODE
end
```

The second phase is to select the best criterion among them and to refine it as a pixel-classifier. A few regions are provided by the user with labels of cells and background; $\mathcal{FG}$ and $\mathcal{BG}$ respectively. The best criterion is selected among all criteria M_k concerning the provided information. The selected criterion is associated with the provided information to define a pixel classifier $\mathcal{L}_{\text{pixel}}$. If the classifier $\mathcal{L}_{\text{pixel}}$ is ambiguous, the user is requested to input more regions to refine the classifier. The details of the selection and the refinement are described later in this section. The system repeats the second phase until the pixel-classifier $\mathcal{L}_{\text{pixel}}$ is sufficient.

The recognition process simply consists of pixel-classification, value assignment, and binarization. The pixel-classes are determined for every pixels by the pixel-classifier $\mathcal{L}_{\text{pixel}}$. Values are assigned to every pixels according to their pixel-classes, and the cell-likeliness map is generated by smoothing operation. Finally, the rough regions of cells are extracted for each input image, by binarization on the map using a given threshold.

Selecting the Optimum Criterion. The selection of the optimum criterion is processed as Algorithm 4. Four k-dimensional vectors are determined for each criterion M_k, based on cumulative count of pixels with some conditions. Each condition and component-wise definition is shown in Algorithm 4; certainty, foreground-likelihood, background-likelihood, and cell-likelihood. The optimum criterion is selected so that it maximizes the difference between foreground-likelihood and the

---

Algorithm 3. Training Process in Drafting Unit

```
void Drafting_Learn( List of Image images )
  features = List of Vector
  foreach Image I_org ∈ images
    I_nrm = Normalize( I_org )
    samples = CalculatePixelFeatures( I_nrm ).Sampling(@samplesPerImage)
    features.Add( samples )
  criteria = List of PixelClassCriterion
  foreach Integer k in @CandidatesOfK
    M_k = KMeans( features, k ).ToPixelClassCriterion()
    criteria.Add( M_k )
  L_pixel = PixelClassifier
  ( FG,BG ) = Region : Empty
  while( not enough )
    RequestUserInput( R_query : if exists )
    L_pixel = SelectOptimumCriterion( criteria )
    R_query = FindAmbiguousAreas( images )
end
```

---

background-likelihood. The difference can be computed by Jensen-Shannon divergence, for example.

Refining the Pixel-Classifier. The pixel-classifier $\mathcal{L}_{pixel}$ consists of classwise associations for certainty, foreground-likelihood, background-likelihood, and cell-likelihood. The certainty value is low when there are few labeled pixels of the class. The system detects ambiguous pixel-classes by checking whether it is associated with low certainty value. If any pixel-class is ambiguous, the system requests more regions to the user. The request involves a specified area R_{query} where the desirable information would be obtained. The area R_{query} is specified from among all the images so that it contains more unlabeled pixels of the desired class.

2.5   Reformation Unit

In this unit, the system improves the rough cell regions which are detected in the drafting unit. The region shape is iteratively improved by a modifier which is acquired by supervised learning. Since the modifier itself is also updated multiple times, it should support stream-learning, e.g., Refs. [7,8]. The training process consists of two phases as shown in Algorithm 5.

The first phase is to update the modifier using all available data, and to obtain the results of modification. The training data for the modifier $\mathcal{L}_{modify}$ is prepared using pixels which are labeled by the user. To reduce complexity of the feature space, these pixels are restricted onto the boundaries of the rough regions. The input part of each training datum is a boundary-feature which is computed from the area around each pixel of the boundary. The output part is a binary value for the label. The boundary-feature should be defined to discriminate efficiently whether the target pixel is inside or outside of the cell. After the modifier $\mathcal{L}_{modify}$

Algorithm 4. Selecting Optimum Criterion in Drafting Unit

```
PixelClassifier SelectOptimumCriterion(List of PixelCriterion criteria)
  classifiers = List of PixelClassifier
  foreach PixelCriterion M_k ∈ criteria
    C_k = PixelClassifier : ( Criterion = M_k )
    ( N_fg , N_bg , N_known , N_all ) = Vector : ( Size = M_k.ClusterCount )
    foreach Image I_org tied to ( FG or BG )
      I_class = PixelClassMap( I_org , M_k )
      foreach Index c in M_k.ClusterCount
        N_fg[c]  += CountOf( I_class[p] = c,  p ∈ FG )
        N_bg[c]  += CountOf( I_class[p] = c,  p ∈ BG )
        N_ans[c] += CountOf( I_class[p] = c,  p ∈ (BG ∪ FG) )
        N_all[c] += CountOf( I_class[p] = c,  p ∈ (Positions of I_org) )
    C_k.Certainty = N_ans ./ N_all
    C_k.ForegroundLikelihood = C_k.Certainty .* ( N_fg ./ N_ans )
    C_k.BackgroundLikelihood = C_k.Certainty .* ( N_bg ./ N_ans )
    C_k.CellLikelihood = C_k.ForegroundLikelihood - C_k.BackgroundLikelihood
    C_k.Score = JS_Divergence(C_k.ForegroundLikelihood, C_k.BackgroundLikelihood)
    classifiers.Add( C_k )
  return arg max C_t.Score,  C_t ∈ classifiers
           C_t
end
```

---

is trained, the region is updated by the modifier. Since the boundary of the region changes accordingly, new training data can be created in the same way reflecting the new states of regions. The additional training is performed on the modifier $\mathcal{L}_{modify}$, and the region is modified again by that. By repeating these procedures for appropriate times, the modifier $\mathcal{L}_{modify}$ can be trained with effective use of all available data.

The second phase is to check whether the modifier is working correctly, and to add some user inputs on the area where the error is noticeable. The final regions of the first phase are displayed to the user for his/her check. If there are any area where the modifier does not work correctly, the user corrects that by giving some regions labeled for cells and background. The system repeats the process from the first phase again with the enriched informations.

The recognition process repeats the modification until some termination condition is satisfied. The termination condition can be set based on the number of repetition or the occurrence rate of region change.

3   An Experiment of Rabbit Chondrocytes

3.1   Dataset and the Typical Difficulties

We applied the proposed system to an image-set of rabbit chondrocytes. The set consists of 16 images by a phase contrast microscope, and each image is 512×512 gray-scaled pixels. Two kinds of cell states are artificially created in order to reproduce the typical difficulties in cell detection. The appearance of

---

Algorithm 5. Training Process in Reformation Unit

```
EdgeModifier Reformation_Learn( List of Region regions )
  𝓛modify = BoundaryModifier
  while( user not satisfied )
    repeat @learnCount_modifiy times
      trainingData = List of Vector
      foreach Region R_i ∈ regions
        E_i = BoundaryOf( R_i )
        foreach Position p ∈ E_i
          BoundaryFeature F_p = CalculateBoundaryFeature( p, R_i )
          if 𝓕𝓖contains p
            then trainingData.Add( F_p, true )
          if 𝓑𝓖contains p
            then trainingData.Add( F_p, false )
      𝓛modify.Learn( training_data )
      regions = Reformation_Recognition( regions )
    if( User_CheckAndCorrect() )
      regions.Clear()
      foreach Image I_j tied to 𝓕𝓖or 𝓑𝓖
        regions.Add( Drafting_Recognition( I_j ) )
end
```

---

the cell is compact in one state, and spreading in the other. Both types are found in Fig. 3. The typical difficulties can be found there as follows:

- The image has uneven brightness over the image.
- The background area does not appear flat.
- The cell area is a mixture of parts with various brightnesses.
- There are several types of appearance of cells.

Most difficulties are due to unavoidable factors, such as out-of-focus dusts, distortion of the petri dish, internal cell structure, cell state and halo effect of phase contrast microscope. They cause the large variation of appearance of cells and background. It prevents both approaches of the manual coding and machine learning. Moreover, it is impossible to pre-classify images into appropriate numbers in a normal situation, because the appearance may change unpredictably even within a single experiment.

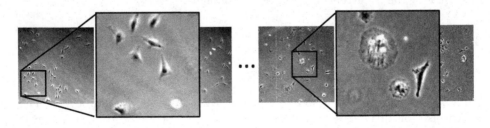

Fig. 3. A typical set of images containing different appearances of cells

3.2 Definitions of Feature Vectors

The feature vectors in each unit can be defined freely, as described before. However, the definitions of the feature vectors can affect the whole performance of the proposed system. Although many good definitions have been proposed, for example [11,12], we used rather simple definitions in the experiment. This is to confirm the performance derived from the structure of the system, excluding the influence of the excellence of the feature definitions. The image-feature and pixel-feature are based on the well-known image processing methods such as average, standard deviation, gradient, and their histograms as shown in Fig. 4. The boundary-feature has a slightly more complex definition to describe enough about the boundary area, as shown in Fig. 5. The boundary-feature consists of two parts, a simple part and a contrastive part. The simple part describes about the central pixel of the boundary area, similar to the pixel-feature. The contrastive part represents difference between inside and outside of a cell. The boundary area is divided into two areas by the edge direction of the central pixel, and sub-features are computed for each of them. Each sub-feature contains the histogram-based components and an additional component which reflects the state of current region. The difference between the sub-features is defined as the

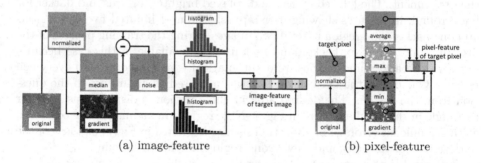

(a) image-feature   (b) pixel-feature

Fig. 4. Definitions of the image-feature and the pixel-feature

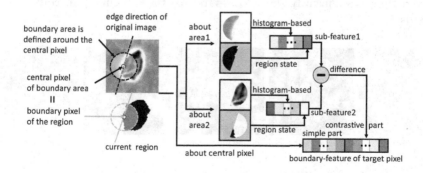

Fig. 5. Definition of the boundary-feature

contrastive part. The boundary-feature of the same position can change every time after the modification due to the additional component.

3.3   Results of the Experiment

The image-set was divided into two subsets of eight images for training and test. Each subset consists of six images under the same condition, one image at different magnification, and one image in different cell-state. It is expected that the system prepare an image-class for six images, and exclude the rest two images as error. Note that a new image-class would not be prepared for the image in different cell-state, because multiple images were required to prepare new image-class in our implementation. In the training process, the system successfully classified six images into a class, and excluded the rest two images. However, in the test-subset, the system failed to exclude a different cell-stated image. It was classified into a same class as six images.

One of the results of the test-subset is shown as Fig. 6. The edges of the regions are overlaid on the image for showing the detected regions in the rightmost picture. The three pictures in the middle indicate that the rough regions were detected with reasonable accuracy and the modification worked well. Figures 7 and 8 shows several examples of the user interactions which occurred in the experiment. The labeled areas are displayed brighter for cells and darker for background, in pictures showing user inputs. The user inputed five cell regions and one background region in total two images during the training process of the drafting unit. Total six background regions were available in this case because fixed-width background regions were automatically generated around every cell regions. One out of the five cell regions was acquired as an answer to the question from the system. The desired area in the question is displayed as brighter rectangle in the middle picture in Fig. 7. Nine regions in an image were corrected as the significant errors of modification results as shown in Fig. 8, e.g. loss of the dark details of the cells, and tiny wrong-regions in the background.

The total time spent on the training process including user interactions was about 20 min. The system required about 30 seconds for processing each image in the test-subset. The occurrence count of user interactions is modest and the required time is sufficiently short, compared to those assumed in conventional methods.

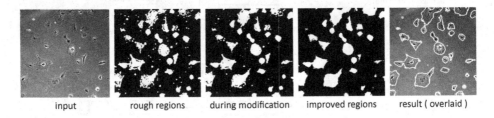

input   rough regions   during modification   improved regions   result (overlaid)

Fig. 6. An example of the results

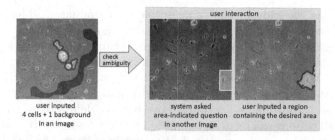

Fig. 7. A example of the interaction in drafting unit

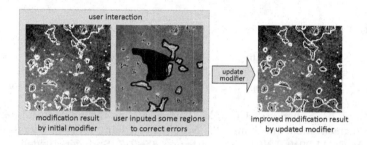

Fig. 8. An example of the interaction in reformation unit

4   Conclusion

We proposed a flexible and efficient cell detection system by combining machine learning and image processing. We verified its effectiveness through an experiment of rabbit chondrocyte images. The proposed system solves some typical problems of cell detection as follows:

It is not efficient to specialize a method for every type of images:
    The system is universally applicable to any type of cell image, because it only consists of basic methods of image processing. The code of the system does not need to be modified, because the behavior of the system are adaptively tuned by supervised machine learning. Parameter collision does not occur, because the images are classified and passed to the appropriately tuned detector.

It is difficult to prepare many training data:
    The system compensates the lack of the training data by associating clustering results and user inputs. It has a function to request the data which effectively improve the detection criterion. The complexity of the feature space is reduced by constraining the targets onto the boundaries of the cell regions.

The system has an advantage that it can quickly and easily prepare a tuned cell detection method for arbitrary sets of images.

In the future, we will consider some problems of evaluation. The behavior of each unit should be evaluated independently with more large dataset. Especially the assortment unit is not evaluated enough because the experiment did not contain multiple valid image-classes. We mainly confirmed efficiency of the system with simple implementation. However, it is worth considering the problems of accuracy and calculation cost to improve practicality of the system. Input images can be classified more appropriately when the definition of image-feature is revised. Cell regions can be detected more precisely and quickly when the definitions of pixel-feature and boundary-feature are improved. The recently revised DNN-related techniques such as [5,6] can be applied to the system to improve its performance.

References

1. Li, K., Kanade, T.: Nonnegative mixed-norm preconditioning for microscopy image segmentation. In: Prince, J.L., Pham, D.L., Myers, K.J. (eds.) IPMI 2009. LNCS, vol. 5636, pp. 362–373. Springer, Heidelberg (2009). doi:10.1007/978-3-642-02498-6_30
2. Yin, Z., Li, K., Kanade, T., Chen, M.: Understanding the optics to aid microscopy image segmentation. In: Jiang, T., Navab, N., Pluim, J.P.W., Viergever, M.A. (eds.) MICCAI 2010. LNCS, vol. 6361, pp. 209–217. Springer, Heidelberg (2010). doi:10.1007/978-3-642-15705-9_26
3. Chen, T., Chefd'hotel, C.: Deep learning based automatic immune cell detection for immunohistochemistry images. In: Wu, G., Zhang, D., Zhou, L. (eds.) MLMI 2014. LNCS, vol. 8679, pp. 17–24. Springer, Cham (2014). doi:10.1007/978-3-319-10581-9_3
4. Dong, B., Shao, L., Da Costa, M., Bandmann, O., Frangi, A.: Deep learning for automatic cell detection in wide-field microscopy Zebrafish images. In: Proceedings of the 12th IEEE International Symposium on Biomedical Imaging, pp. 772–776 (2015)
5. Li, Y., Paluri, M., Rehg, J.M., Dollár, P.: Unsupervised learning of edges. In: Proceedings of the IEEE Conference on Computer Vision and Pattern Recognition, pp. 1619–1627 (2016)
6. Vinyals, O., Blundell, C., Lillicrap, T., Wierstra, D.: Matching networks for one shot learning. In: Advances in Neural Information Processing Systems, pp. 3630–3638 (2016)
7. Crammer, K., Kulesza, A., Dredze, M.: Adaptive regularization of weight vectors. In: Advances in Neural Information Processing Systems, pp. 414–422 (2009)
8. Wang, J., Zhao, P., Hoi, S.C.: Exact soft confidence-weighted learning. In: Proceedings of the 29th International Conference on Machine Learning (2012)
9. Serra, J.: Image Analysis and Mathematical Morphology, vol. 1. Academic press, London (1982)
10. Canny, J.: A computational approach to edge detection. IEEE Trans. Pattern Anal. Mach. Intell. **6**, 679–698 (1986)
11. Lowe, D.G.: Object recognition from local scale-invariant features. In: Proceedings of the 7th IEEE International Conference on Computer Vision, vol. 2, pp. 1150–1157 (1999)
12. Dalal, N., Triggs, B.: Histograms of oriented gradients for human detection. In: Proceedings of the IEEE Conference on Computer Vision and Pattern Recognition, vol. 1, pp. 886–893 (2005)

A Fuzzy Colour Model Sensitive to the Context: Study Cases Using PRAGR and Logics

Zoe Falomir[1]([✉]) and Luis Gonzalez-Abril[2]

[1] University of Bremen, Enrique-Schmidt-Str. 5, Bremen, Germany
zfalomir@uni-bremen.de
[2] Universidad de Sevilla, Avd. Ramon y Cajal, 1, Sevilla, Spain
luisgon@us.es

Abstract. A fuzzy colour model is defined to deal with human-machine communication situations where perceptual and conceptual deviations can appear. Logics have been defined to combine this model with the Probabilistic Reference And GRounding mechanism (PRAGR) (Mast and Wolter 2013) in order to obtain the most acceptable and appropriate colour descriptor depending on the situation. Two case studies are presented and promising results are obtained.

1 Introduction

Human-machine understanding is crucial for modeling cognitive decisions for AI. Misunderstandings usually appear when entities are classified differently by the interactuators due to distinct perceptions or due to conceptual deviation. A cognitive system must evaluate which category within a domain to assign (e.g., for colours *yellow* or *orange*) in order to maximise the chances of the listener to identify the correct target object in the given context.

The challenge of colour categorization is tackled in this paper. Colour perception is very subjective since there are colours which may be named differently depending on the person who refers to them. There are other cognitive effects in colour perception, such as sensitivity to brightness or/and colour context (Lotto and Purves 1999), which may cause conceptual mismatch and communication failure between humans and machines. Examples of these effects are shown in Fig. 1. There is also empirical evidence which confirms that object colour perception changes with change in illumination and background (Helson 1938).

Spranger and Pauw (2012) show that overlapping, graded colour categories can improve the chances of communicative success under circumstances of perceptual and conceptual deviation. Furthermore, it is well-known that the concept of *fuzzy set* provides a methodology for translating the numerical data obtained from the world into linguistic categories with a degree of believing which can be given a meaning and used for reasoning (Zadeh 1965). Thus, this is an attractive methodology to model colours since providing a fuzzy model for colour naming may allow the adaptation of the system's usage of colour terms to the idiosyncrasies of user groups or individuals.

© Springer International Publishing AG 2017
V. Torra et al. (Eds.): MDAI 2017, LNAI 10571, pp. 207–219, 2017.
DOI: 10.1007/978-3-319-67422-3_18

(a)   (b)   (c)

Fig. 1. (a) Example of Perceptual deviation: is it a green or a yellow square? Other examples of contextual sensitivity of brightness perception are also shown: the colour of the small circle is the same in both objects, in both scenes (b) colour with no hue, and (c) colour with hue. (Color figure online)

In the literature, fuzzy colour descriptors were defined on colour models, as for example: (i) the approach for computational colour categorisation and naming based on the CIE Lab colour space and fuzzy partitioning formulated by Menegaz et al. (2007); (ii) the fuzzy colour categories on the Musell Colour Solid and the HCL colour space and similarity values based on the Fuzzy C-Means by Seaborn et al. (2005); (iii) the approach to automatically design customised fuzzy colour spaces on any Euclidean crisp space proposed by Soto-Hidalgo et al. (2010), etc.

All these previous works provide evidence for the effectiveness of applying fuzzy methods for colour naming. However, in the literature, very few approaches manage the context for colour naming. Meo et al. (2014) use uncertain boundaries of colour terms in their system, which can generate appropriate colour names in referential situations for discriminating similar colour patches. Mast et al. (2016) combined the Probabilistic Reference And GRounding (PRAGR) mechanism with a vague qualitative colour model in a linguistic environment based on ontologies and the open world assumption (OWA).

The main contribution in this paper is the definition of a fuzzy qualitative colour model (Fuzzy-QCD) using first-order logics and the closed world assumption (CWA). In order to deal with context perception, the Fuzzy-QCD model is combined with PRAGR using the logics defined.

The remainder of this paper is organized as follows. A Fuzzy Colour Descriptor is presented in Sect. 2 and its membership functions are given in Sect. 3. Section 4 provides logics for the Fuzzy-QCD. Section 5 combines these logics with PRAGR. Sections 5 and 6 provides two case studies used in the implementation of these logics which prove applicability of the Fuzzy-QCD+PRAGR approach. Finally a discussion is provided.

2   A Fuzzy Qualitative Colour Descriptor (Fuzzy-QCD)

Let us define a Reference System (RS) for our Fuzzy Qualitative Colour Descriptor (QCD) as:

$$\text{Fuzzy-QCRS} = \{uH, uS, uL, C_{FzSet_{1..5}}\}$$

where uH, uS and uL are the unit of Hue, Saturation and Lightness, respectively, in the HSL colour space (Fig. 2); $C_{FzSet_{1..5}}$ refers to the selected colour names and the fuzzy sets related to them, as follows:

- $C_{FzSet_1} = \{(G_1, \mu_{G_1}), ..., (G_i, \mu_{G_i}), ..., (G_\ell, \mu_{G_\ell})\}$ where ℓ colour names are defined for the grey scale in C_{FzSet_1} by fuzzy sets.

- $C_{FzSet_2} = \{(R_1, \mu_{R_1}), ..., (R_j, \mu_{R_j}), ..., (R_r, \mu_{R_r})\}$ where r colour names are defined for the rainbow scale in C_{FzSet_2} and are considered the more saturated or stronger ones. The saturation of the R_i colours take values between $r_{us_{MIN}}$ and 100, whereas their luminance take values between $r_{ul_{MIN}}$ and $r_{ul_{MAX}}$. Thus, the different values of hue (r_{uh_r}) take values between 0 and 360 and determine the colour names defined for the C_{FzSet_2} set.

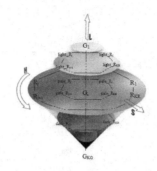

Fig. 2. The Fuzzy Qualitative Colour Descriptor (Fuzzy-QCD) build on the HSL colour space. (Color figure online)

- $C_{FzSet_3} = \{pale\text{-} + C_{FzSet_2}\}$ where r pale colour names are defined in C_{FzSet_3} by adding the prefix *pale-* to the colours defined for the rainbow scale, C_{FzSet_2}. The colour names defined in C_{FzSet_3} have the same interval values of hue as rainbow colours (C_{FzSet_2}). The lightness intervals also coincide, but they differ from rainbow colours in their saturation, which can take values between $g_{us_{MAX}}$ and $r_{us_{MIN}}$.

- $C_{FzSet_4} = \{light\text{-} + C_{FzSet_2}\}$ and $C_{FzSet_5} = \{dark\text{-} + C_{FzSet_2}\}$ where r light and dark colour names are defined in C_{FzSet_4} and C_{FzSet_5}, respectively, by addition of the prefixes *dark-* and *light-* to the colour names in the rainbow scale (C_{FzSet_2}). The intervals of values for dark and light colour sets, C_{FzSet_4} and C_{FzSet_5}, take the same values of hue as rainbow colours in C_{FzSet_2}, respectively. The saturation intervals also coincide, but the lightness coordinate (uL) differs and determines the luminosity of the colour (dark or light) taking values between $r_{ul_{MAX}}$ and 100 for light colours and between r_{ul} and $r_{ul_{MIN}}$ for dark colours.

It is worth noting that the parameters ℓ and r depend on the granularity that an expert defines for a scenario. The rest of parameters, $r_{us_{MIN}}$, $r_{ul_{MIN}}$, $r_{ul_{MAX}}$, $g_{us_{MAX}}$, $r_{us_{MIN}}$ and r_{ul} can be defined by experimentation. The Fuzzy-QCD is parameterised as a baseline using data of a collection of colour intervals (each one assumed to be fully representative of a certain colour term) coming from previous experiments (Falomir et al. 2015). However, those crisp intervals have been adapted to a more intuitive definition where values are overlapped. Colours in the grey scale, C_{FzSet_1}, have uncertain boundaries in the dimension of lightness, such as the prototypical colours, C_{FzSet_2}, in the dimension of hue. Moreover, prototypical colours are defined in the whole range of lightness, because a *light-red* or a *dark-red colour* is considered also *red*. As a result, the intervals defining *light/dark* colours are overlapping also prototypical colours. Table 1 shows the values used for parameterisation of the Fuzzy-QCD extracted from the experimentation aforementioned.

Table 1. HSL overlapping intervals for colour names.

	Colour name	H(h_0, h_1)	S(s_0, s_1)	L(l_0, l_1)
C_{FzSet_1}	black	[0, 360]	[0, 20]	[0, 20)
	dark-grey			[20, 30)
	grey			[30, 50)
	light-grey			[50, 75)
	white			[75, 100]
C_{FzSet_2}	red	(335, 360] $\wedge$ [0, 20]	(50, 100]	(20, 80]
	orange	(20, 50]		
	yellow	(50, 80]		
	green	(80, 160]		
	turquoise	(160, 200]		
	blue	(200, 260]		
	purple	(260, 300]		
	pink	(300, 335]		
C_{FzSet_3}	pale + C_{FzSet_2}	Idem	(20, 50]	(40, 100]
C_{FzSet_4}	light + C_{FzSet_2}	Idem	(50, 100]	(55, 100]
C_{FzSet_5}	dark + C_{FzSet_2}	Idem	(50, 100]	(0, 40]

Moreover, inspired by other works in the literature (Palmer and Schloss 2010), this Fuzzy-QCD model has also defined some equivalent colours such as: *dark-orange* $\equiv$ *brown*, *dark-yellow* $\equiv$ *olive*, *light-red* $\equiv$ *pastel-pink* according to the Inter-Society Colour Council - National Bureau of Standards (ISCC-NBS[1]).

3 Fuzzy-QCD Membership Functions

Let us define a fuzzy degree of believing for each colour defined on HSL internals: $[h_0, h_1] \times [s_0, s_1] \times [l_0, l_1] \equiv B_r(h, s) \times B_{lr}(lc)$[2].

Thus, a colour A in the Fuzzy-QCD is determined by the parameters $(h_c, h_r, s_c, s_r, l_c, l_r)$ where (h_c, s_c, l_c) and (h_r, s_r, l_r) are the centroids and the half-amplitudes of the intervals in hue (h_0, h_1), saturation (s_0, s_1) and lightness (l_0, l_1) for each defined colour, respectively.

The membership function of A, $\mu_A : [0.360] \times [0, 100] \times [0, 100] \rightarrow [0, 1]$ is defined from a three-dimensional Radial Basis Function (RBF) as follows:

$$\mu_A(h, s, l) = e^{-\frac{1}{2}((\frac{h-h_c}{h_r})^2 + (\frac{s-s_c}{s_r})^2 + (\frac{l-l_c}{l_r})^2)} \tag{1}$$

---

[1] ISCC-NBS: http://tx4.us/nbs-iscc.htm (Accessed June 2017).

[2] Note that, given an open interval (analogously for another kind of interval) of finite dimension, there are two main ways to represent it: from the extreme points as *(a, b)* (classical notation) or as an open ball B$_r$(c) (Borelian notation) where $c = (a+b)/2$ (centre) and $r = (b-a)/2$ (radius).

Note that, from the spatial structure of the HSL colour space, it is obtained that:

- In C_{FzSet_1}, the interval values obtained in HSL coordinates correspond to cylinders in the Cartesian coordinate system. Therefore, for the colours in the grey scale, it is holds that:
  (h_c, s_c, l_c) where $h_c = s_c = s_0 = min_{saturation}$ and $l_c = \frac{l_0+l_1}{2}$
  (h_r, s_r, l_r) where $h_r = s_r = s_1 = max_{saturation}$ and $l_r = \frac{l_1-l_0}{2}$
  As $h_c = s_c$ and $h_r = s_r$, then (1) is independent of hue and it can be rewritten as:
  $$\mu(h, s, l) = e^{-\frac{1}{2}((\frac{s-s_c}{s_r})^2+(\frac{l-l_c}{l_r})^2)} \tag{2}$$

Note that for the colour *black*, $l_c = l_0 = 0$ and for the colour *white*, $l_c = l_1 = 100$.

- In all other cases except prototypical colours (C_{FzSet_2}), the interval values in HSL correspond to wedges in the Cartesian axis (see Fig. 3). Therefore, for the colours in the rest of scales, it is holds that:
  (h_c, s_c, l_c) where $h_c = \frac{h_0+h_1}{2}$, $s_c = \sqrt{\frac{s_0^2+s_1^2}{2}}$; and $l_c = \frac{l_0+l_1}{2}$
  (h_r, s_r, l_r) where $h_r = \frac{h_1-h_0}{2}$, $s_r = \sqrt{\frac{s_0^2+s_1^2}{2}}$ that is, $s_c = s_r$, and $l_r = \frac{l_1-l_0}{2}$
  The exception is the red colour, which is divided into two parts in order to cover the starting and ending part of the central circle in HSL. For the first, $h_c = h_0 = 0$ and $h_r = h_1 - h_0$, whereas for the second *red*, $h_c = h_1 = 360$ and $h_r = h_1 - h_0$.

- For the prototypical colours (C_{FzSet_2}), setting $s_c = s_1 = 100$ means that a maximally saturated colour gets the highest certainty, and since reducing saturation reduces certainty.

A colour described in HSL coordinates can get more than one colour name with a different degrees of believing (calculated from 1 and 2). Thus, in order to normalise these values, the following operation is applied:

$$\mu_A^N(h, s, l) = \frac{\mu_A(h, s, l)}{\sum\limits_{A_i \text{ of the Fuzzy-QCD} } \mu_{A_i}(h, s, l)}$$

where the input variables are hue (h), saturation (s) and lightness (l) as the coordinates of the colour in HSL and $\sum \mu(h, s, l)$ is the total amount of certainties obtained for these HSL colour coordinates.

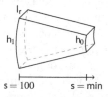

Fig. 3. Wedges which determine the colours in the most saturated colour scale C_{FzSet_2}, the pale scale C_{FzSet_3}, the light scale C_{FzSet_4} and the dark scale C_{FzSet_5}.

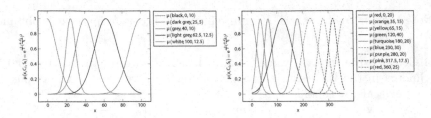

Fig. 4. Fuzzy colours: (a) in the grey scale in the lightness dimension (L); (b) in the prototypical scale in the hue dimension (H). (Color figure online)

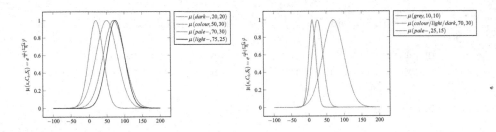

Fig. 5. Fuzzy colours in: (a) the dimension of lightness (L) and (b) the dimension of saturation (S). (Color figure online)

The HSL intervals in Table 1 are used to obtain the parameters (h_c, s_c, l_c) and (h_r, s_r, l_r) as aforementioned. These parameters are applied to (1) to generate three dimensional RBFs for obtaining the degree of believing for each colour name.

Some RBF functions in one-dimension are represented here. Figure 4(a) shows the membership functions of the colours in the grey scale, C_{FzSet_1}, taking into account the dimension of lightness.

The membership functions of the prototypical colours, C_{FzSet_2}, in the dimension of hue are depicted in Fig. 4(b). Figure 5(a) shows the membership function of a prototypical colour, *red*, defined in the whole range of lightness, since *light-red* and *dark-red colour* are considered also *red*.

Figure 5(b) shows the membership functions obtained for any prototypical colour defined in the dimension of lightness or saturation, respectively. Note how the membership functions overlap as a result of the RBF parameters.

Finally, Fig. 6 shows a two-dimensional RBF membership function of a prototypical colour (*orange*) depending on the amount of hue and the amount of saturation, considered together.

4   Logics for the Fuzzy-QCD

The Fuzzy-QCD presented is used for generating logic descriptions of colours related to objects obtained in a digital image (Falomir 2017):

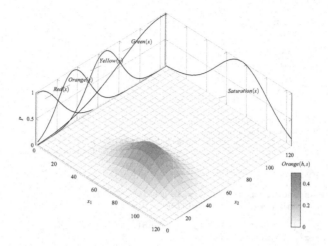

Fig. 6. Radial Basis Function in two dimensions (hue and saturation) for the colour *orange*. (Color figure online)

```
hasQSD_category(SceneID,ObjectID,ShapeCategory,Certainty).
hasHSL(SceneID,ObjectID,H,S,L).
has_FuzzyQCD(ObjectID, Fuzzy-QC_LAB1..5).
```

where *ObjectID* refers to the object identifier in the image, *hasHSL* refers to the colour coordinates in the HSL colour space and $Fuzzy\text{-}QC_{LAB1..5}$ refers to the colour names and their certainties given an object in an image. Figure 7 shows an example of a digital image where the qualitative shape description (QSD*) (Falomir and Olteţeanu 2015) of the objects is extracted together with its Fuzzy-QCD.

5   The Probabilistic Reference And GRounding Mechanism (PRAGR) Combined with Fuzzy-QCD Using Logics

The Probabilistic Reference And GRounding mechanism (PRAGR) (Mast and Wolter 2013) can determine, in a given situation, the most appropriate referring expression to use for describing a particular object. PRAGR is based on: *acceptability*, *discriminatory power* and *appropriateness*. This section describes PRAGR using logics[3]. The results shown are related to the scene in Fig. 7.

The *acceptability* of a description D is the conditional probability $P(D|x)$ that the listener will accept D as a good description for object x. For simple properties, like colour terms, *acceptability* can correspond to the fuzzy degree of believing obtained by the Fuzzy-QCD presented here. Using logic definitions

---

[3] A Prolog program has been developed for connecting Fuzzy-QCD and PRAGR and it is available for download: https://sites.google.com/site/cogqda/publications.

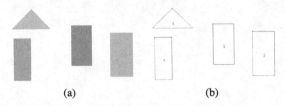

(a)   (b)

```
hasQSD_category(scene-z1, object-1, triangle, 1.0).
hasHSL(scene-z1, object-1, 80,32,67).
has_FuzzyQCD(scene-z1, object-1, [[pale-yellow, 0.6],[pale-green, 0.58],
[yellow, 0.47],[green, 0.51],[light-yellow, 0.5],[light-green, 0.54]]).

hasQSD_category(scene-z1, object-2, rectangle, 1.0).
hasHSL(scene-z1, object-2, 94,20,50).
has_FuzzyQCD(scene-z1,object-2, [[pale-yellow, 0.12],
[pale-green, 0.65],[yellow, 0.13],[green, 0.77]]).

hasQSD_category(scene-z1, object-3, rectangle, 1.0).
hasHSL(scene-z1, object-3, 63,3,72).
has_FuzzyQCD(scene-z1, object-3, [[light-grey, 0.7],[white, 0.065]]).

hasQSD_category(scene-z1, object-4, rectangle, 1.0).
hasHSL(scene-z1, object-4, 142,23,63).
has_FuzzyQCD(scene-z1, object-4, [[pale-green, 0.84],[pale-turquoise, 0.06],
[green, 0.75],[turquoise, 0.082],[light-green, 0.67],
[light-turquoise, 0.074]]).
```

Fig. 7. Example of Fuzzy-QCDs obtained for the objects in the image. (Color figure online)

based on the previous facts showed by Fig. 7, acceptabilities of colours (i.e. green) can be extracted:

```
acceptability_colour(Scene,Obj,Colour,C):-
  has_FuzzyQCD(Scene,Obj,ColourList),
  extract_certainty(Colour,ColourList,C).
```

```
?- acceptability_colour(scene-z1,
   Obj,green,Certainty).
Obj = object-1, Certainty = 0.51 ;
Obj = object-2, Certainty = 0.77 ;
Obj = object-3, Certainty = 0 ;
Obj = object-4, Certainty = 0.75
```

The *discriminatory power* (DP) is the power of a description to discriminate the intended object from the rest. It can be defined as the conditional probability $P(x|D)$ of determining the correct object, given a description. DP and acceptability are thus interrelated by the Theorem of Bayes: $P(x|D) = \frac{P(D|x)P(x)}{P(D)}$ where $P(x)$ is the prior probability of selecting object x ($P(x) = \frac{1}{N}$ where N is the total number of objects in the scene, assuming equal probability for selecting any one of them – maximum entropy) and $P(D)$ is the prior probability of the description. Using logic definitions and the facts from Fig. 7, prior probabilities of selecting a colour (i.e. yellow, green) are obtained as follows:

```
prob_colour_in_a_scene(Scene,Colour,PD):-
   objects_in_scene(Scene,List,N),
   coloured_objects_in_scene(Scene,Colour,List,Accumulative),
   PD is Accumulative/N.
?- prob_colour_in_a_scene(scene-z1,green,PD). PD = 0.5075
```

DP compares the acceptability of D for the target object to the acceptability of D for other distractors. For identifying the most likely referent of a given description, PRAGR selects the object for which D has the highest *acceptability*. Logically, the DP of a colour for a particular object can be computed from the previous facts and definitions as follows:

```
discriminatory_power(Scene, Object, Colour, DP):-
   acceptability_colour(Scene,Object,Colour,PDx),
   prob_choosing_object(Scene,Px),
   prob_colour_in_a_scene(Scene,Colour,PD),
   DP is PDx*Px/PD.

?- discriminatory_power(scene-z1,object-1,green, DP). DP = 0.25 .
?- discriminatory_power(scene-z1,object-1,yellow, DP). DP = 0.783 .
```

When generating referring expressions, PRAGR jointly maximises *acceptability* and *DP*, selecting an expression that yields the best balance between providing an adequate description of an object per se, and a description that distinguishes it from the other objects in the context, that is, it obtains the more appropriate description depending on the context. In order to maximize the *appropriateness* of the description, a parameter $\alpha \in [0, 1]$ is introduced: $D_x^* := \arg\max_D (1 - \alpha)P(x|D) + \alpha P(D|x)$.

The parameter α determines the weighting of *acceptability*, with a value of 0 indicating that *acceptability* is ignored, and a value of 1 meaning that only *acceptability* is considered, and DP is ignored. Previous experimentations by Mast et al. (2016), identified that a choice of $\alpha \in [0.1, 0.4]$ leads to intuitive descriptions. Using logics, the appropriateness of a colour (i.e. green, yellow) for an object can be obtained as follows:

```
appropriateness(Scene, Object, Colour, Alfa, Dx):-
   discriminatory_power(Scene, Object, Colour, PxD),
   acceptability_colour(Scene,Object,Colour,PDx),
   Dx is (1-Alfa)*PxD + Alfa*PDx.

?- appropriateness(scene-z1, object-1, yellow, 0.4, Dx). Dx = 0.658 .
?- appropriateness(scene-z1, object-1, green, 0.4, Dx). Dx = 0.355 .
```

Table 2 presents a summary of the values obtained by Fuzzy-QCD+PRAGR using logics and the facts from the scene in Fig. 7. As a result, the more appropriate colour descriptor for object-1 is *light-yellow* ($D_x^* = 0.8$), which is quite acceptable for that particular object ($P(D|x) = 0.5$) but highly discriminative

Table 2. Values of PRAGR for different colour descriptors. Note that $P(x) = 0.25$ for all the cases, since there are 4 objects in the scene.

Scene		Acceptability $P(D|x)$	$P(D)$	Discriminatory Power (DP)	Appropriateness D_x^* $\alpha = 0.4$
Object	Colour				
object-1	green	0.51	0.5075	0.250	0.355
object-1	yellow	0.47	0.1500	0.780	0.658
object-1	light-yellow	0.50	0.1250	1.000	0.800
object-2	green	0.77	0.5075	0.380	0.536
object-2	pale-green	0.65	0.5175	0.314	0.448
object-4	green	0.75	0.5075	0.370	0.520
object-4	pale-green	0.84	0.5175	0.400	0.579
object-4	light-green	0.67	0.3025	0.550	0.600

$(DP = 1.0)$. For object-2, the more appropriate colour is *green* ($D_x^* = 0.536$), and for object-4 is *light-green* ($D_x^* = 0.6$).

6   Flexibility of Fuzzy-QCD + PRAGR Depending on the Context

In this section, another use case for combining Fuzzy-QCD with PRAGR is provided considering the scene examples in Fig. 8. This case is cognitively interesting because the red colour in scene-a is usually perceived more reddish than in the scene-b, where people usually perceive it as more orange. However, in both scenes, the HSL colour coordinates of the objects are nearly the same (HSL[5, 56, 64] vs. HSL[8, 59, 61]).

Table 3 presents the values of *acceptability, discriminatory power* and *appropriateness* calculated by Fuzzy-QCD+PRAGR for object-4 (square located up) in both scenes in Fig. 8. Results show that, in scene-b, the *orange* colour obtains a higher appropriateness value than *red* (using $\alpha = 0.18$). Even though *red* has a higher acceptability value, the potential acceptability of colour *red* for the distractor object-1 ($P(red|object-1) = 0.174$) leads to a low discriminatory power, in favour of the less acceptable, but more discriminating *orange*.

In scene-b, the objects would be referred as the *pink* circle, the *orange* circle, the *orange* square and the *pink* square. Thus, no object is referred as *red* to maximize appropriateness. However, the acceptability obtained by Fuzzy-QCD for *red*, remains in the agent knowledge-base, so that, if *red* is used by a person in a dialogue, this colour would still be understood by the agent.

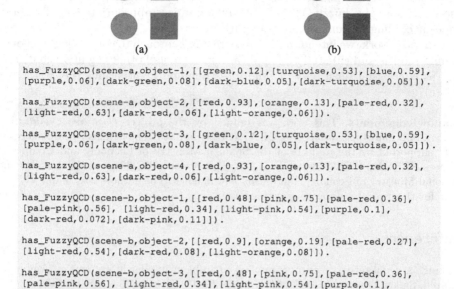

```
has_FuzzyQCD(scene-a,object-1,[[green,0.12],[turquoise,0.53],[blue,0.59],
[purple,0.06],[dark-green,0.08],[dark-blue,0.05],[dark-turquoise,0.05]]).

has_FuzzyQCD(scene-a,object-2,[[red,0.93],[orange,0.13],[pale-red,0.32],
[light-red,0.63],[dark-red,0.06],[light-orange,0.06]]).

has_FuzzyQCD(scene-a,object-3,[[green,0.12],[turquoise,0.53],[blue,0.59],
[purple,0.06],[dark-green,0.08],[dark-blue, 0.05],[dark-turquoise,0.05]]).

has_FuzzyQCD(scene-a,object-4,[[red,0.93],[orange,0.13],[pale-red,0.32],
[light-red,0.63],[dark-red,0.06],[light-orange,0.06]]).

has_FuzzyQCD(scene-b,object-1,[[red,0.48],[pink,0.75],[pale-red,0.36],
[pale-pink,0.56], [light-red,0.34],[light-pink,0.54],[purple,0.1],
[dark-red,0.072],[dark-pink,0.11]]).

has_FuzzyQCD(scene-b,object-2,[[red,0.9],[orange,0.19],[pale-red,0.27],
[light-red,0.54],[dark-red,0.08],[light-orange,0.08]]).

has_FuzzyQCD(scene-b,object-3,[[red,0.48],[pink,0.75],[pale-red,0.36],
[pale-pink,0.56], [light-red,0.34],[light-pink,0.54],[purple,0.1],
[dark-red,0.08], [dark-pink, 0.11]]).

has_FuzzyQCD(scene-b,object-4,[[red,0.9],[orange,0.19],[pale-red,0.27],
[light-red,0.54],[dark-red,0.08],[light-orange, 0.08]]).
```

Fig. 8. Contextual sensitivity of colour perception –note that red is more distinguishing in (a)– and Fuzzy-QCD obtained for the objects. Note that they are numbered anti-clockwise (Color figure online).

Table 3. Evaluation values of Fuzzy-QCD+PRAGR for object-4 in both scenes, where there are 4 objects in each scene, thus $P(x) = 0.25$.

	Object	Colour	Acceptability $P(D|x)$	$P(D)$	Discriminatory Power (DP)	Appropriateness D_x^* $\alpha = 0.18$
Scene-a	object-4	red	0.93	0.465	0.500	0.5774
	object-4	orange	0.13	0.065	0.500	0.4334
Scene-b	object-4	red	0.90	0.690	0.326	0.4290
	object-4	orange	0.19	0.095	0.500	0.4440

7   Discussion

A fuzzy colour model (Fuzzy-QCD) is defined and combined with the Probabilistic Reference And GRounding mechanism (PRAGR) (Mast and Wolter 2013) in order to obtain the most acceptable and appropriate colour descriptor based on cognitive properties and depending on the situation. First order logics have been used to implement the Fuzzy-QCD and Swi-Prolog[4] has been used as the testing platform (Wielemaker et al. 2012). Two case studies are presented and the results obtained by Fuzzy-QCD+PRAGR maximizes appropriateness of colour names according to human understanding.

As future work, we intend to: (i) compare our approach to conventional colour naming models and (ii) study its applicability to model other cognitive contexts such as contextual sensitivity of brightness perception situations (i.e. Fig. 1(b) and (c)).

Acknowledgements. This work was funded by the project *Cognitive Qualitative Descriptions and Applications* (CogQDA) at Universität Bremen. This research is partially supported by the projects of the Spanish Ministry of Economy and Competitiveness HERMES (TIN2013-46801-C4-1-R) and Simon (TIC-8052) of the Andalusian Regional Ministry of Economy, Innovation and Science. The authors also thank Vivien Mast for the use cases appearing in this paper.

References

Falomir, Z.: Qualitative descriptors applied to ambient intelligent systems. J. Ambient Intell. Smart Environ. (JAISE) **9**(1), 21–39 (2017). doi:10.3233/AIS-160418

Falomir, Z., Museros, L., Gonzalez-Abril, L.: A model for colour naming and comparing based on conceptual neighbourhood. An application for comparing art compositions. Knowl. Based Syst. **81**, 1–21 (2015). doi:10.1016/j.knosys.2014.12.013

Falomir, Z., Olteţeanu, A.-M.: Logics based on qualitative descriptors for scene understanding. Neurocomputing **161**, 3–16 (2015). doi:10.1016/j.neucom.2015.01.074

Helson, H.: Fundamental problems in color vision. i. the principle governing changes in hue, saturation, and lightness of non-selective samples in chromatic illumination. J. Exp. Psychol. **23**(5), 439–476 (1938)

Lotto, R., Purves, D.: The effects of color on brightness. Nat. Neurosci. **2**(11), 1010–1014 (1999)

Mast, V., Falomir, Z., Wolter, D.: Probabilistic reference and grounding with PRAGR for dialogues with robots. J. Exp. Theor. Artif. Intell. **28**(5), 889–911 (2016). doi:10.1080/0952813X.2016.1154611

Mast, V., Wolter, D.: A probabilistic framework for object descriptions in indoor route instructions. In: Tenbrink, T., Stell, J., Galton, A., Wood, Z. (eds.) COSIT 2013. LNCS, vol. 8116, pp. 185–204. Springer, Cham (2013). doi:10.1007/978-3-319-01790-7_11

Menegaz, G., Troter, A.L., Sequeira, J., Boi, J.M.: A discrete model for color naming. EURASIP J. Appl. Signal Process **2007**(1), 1–10 (2007). Special Issue on Image Perception

Meo, T., McMahan, B., Stone, M.: Generating and resolving vague color references. In: Rieser, V., Muller, P. (eds.) Proceedings of SemDial 2014/DialWatt, pp. 107–115 (2014)

Palmer, S.E., Schloss, K.B.: An ecological valence theory of human color preference. Proc. Natl. Acad. Sci. **107**(19), 8877–8882 (2010)

---

[4] SWI-Prolog: http://www.swi-prolog.org/.

Seaborn, M., Hepplewhite, L., Stonham, T.J.: Fuzzy colour category map for the measurement of colour similarity and dissimilarity. Pattern Recogn. **38**(2), 165–177 (2005)

Soto-Hidalgo, J.M., Chamorro-Martinez, J., Sanchez, D.: A new approach for defining a fuzzy color space. In: 2010 IEEE International Conference on Fuzzy Systems (FUZZ), pp. 1–6 (2010)

Spranger, M., Pauw, S.: Dealing with perceptual deviation - vague semantics for spatial language and quantification. In: Steels, L., Hild, M. (eds.) Language Grounding in Robots, pp. 173–192. Springer, Heidelberg (2012)

Wielemaker, J., Schrijvers, T., Triska, M., Lager, T.: SWI-Prolog. Theor. Pract. Logic Program. (TPLP) **12**(1–2), 67–96 (2012)

Zadeh, L.: Fuzzy sets. Inf. Control **8**, 338–353 (1965)

Joint Subjective Opinions

Magdalena Ivanovska[1], Audun Jøsang[1(✉)], Jie Zhang[2], and Shuo Chen[2]

[1] University of Oslo, Oslo, Norway
audun.josang@mn.uio.no
[2] Nanyang Technological University, Singapore, Singapore

Abstract. Subjective opinions generalize probability distributions by including degrees of uncertainty which reflect lack of confidence in the probabilities. This paper describes a method for computing the joint subjective opinion of two variables which can be generalized to a method for computing joint subjective opinions over multiple variables in a subjective Bayesian network. We show how the joint opinions can be marginalized to provide subjective opinions on a reduced number of variables. With an example we compare the marginalization of a joint opinion with subjective logic deduction which also produces a marginal opinion.

1 Introduction

In many contexts of probabilistic reasoning one deals with uncertainty and incomplete information. This leads to rough estimation or vague guesses of the probabilities which can influence the correctness of the conclusions in the analysis. Many formalisms for dealing with uncertain probabilities have been proposed in the literature, such as the work on imprecise probabilities [13], which when applied to Bayesian networks produces a certain type of credal networks [1], where the main idea is to work with probability intervals. In contrast to imprecise probabilities, subjective logic [6] is a formalism that offers explicit treatment of the uncertainty about probabilities in both representation and inference. The arguments in subjective logic are *subjective opinions* on random variables. A subjective opinion includes a *belief mass distribution* over the values of the variable, complemented by an *uncertainty mass*, which together reflect a specific analysis of the probability distribution; and a *base rate* probability distribution over the variable, reflecting a prior domain knowledge that is relevant to the specific analysis. In an attempt to convey as reliable as possible inference with uncertain probabilistic information, subjective logic provides many operators that generalize calculations with probability distributions, for a detailed overview see [6].

This paper introduces a method for computing joint subjective opinions on two random variables X and Y based on a subjective opinion on X and conditional opinions on Y given X. The result is a subjective opinion on the joint distribution of X and Y (as opposed to the *belief fusion* operation in subjective logic that fuses different sources' opinions on the same variable). The marginal of a joint opinion should ideally be equal to the deduced subjective opinion on

© Springer International Publishing AG 2017
V. Torra et al. (Eds.): MDAI 2017, LNAI 10571, pp. 220–233, 2017.
DOI: 10.1007/978-3-319-67422-3_19

the same variable. This equality always holds for probability distributions, but produces slightly different uncertainty masses for subjective opinions, since subjective opinions are equivalent to PDFs (probability density functions) and each operation induces an approximation of the corresponding resulting PDF as well. In that sense, the more direct the method of determining the required subjective opinion, the less approximate it is. To illustrate the latter, we present an example which compares the indirect method of marginalization of a joint opinion with the direct method of subjective logic deduction.

The joint operation can be extended to multiple variables, which enables determining the joint opinions in a subjective Bayesian network (SBN) - an extension of a Bayesian network (BN) where the conditional probability distributions are substituted with subjective opinions, for details see [2].

2   Subjective Opinions

Let X be a random variable. A *subjective opinion* on X [4] is a tuple:

$$\omega_X = (\boldsymbol{b}_X, u_X, \boldsymbol{a}_X) , \tag{1}$$

where $\boldsymbol{b}_X : \mathbb{X} \to [0, 1]$ is a *belief mass distribution* over X, $u_X \in [0, 1]$ is an *uncertainty mass*, and $\boldsymbol{a}_X : \mathbb{X} \to [0, 1]$ is a *base rate distribution* over X, satisfying the additivity constraints:[1]

$$u_X + \sum_{x \in \mathbb{X}} \boldsymbol{b}_X(x) = 1 \quad \text{and} \quad \sum_{x \in \mathbb{X}} \boldsymbol{a}_X(x) = 1 . \tag{2}$$

The beliefs and the uncertainty mass are a result of a specific analysis of the random variable by applying expert knowledge, experiments, a personal judgement, etc. $\boldsymbol{b}_X(x)$ is the belief that X takes the value x expressed as a degree in $[0, 1]$. It represents the amount of experimental or analytical evidence in favour of x. u_X is a scalar, representing the degree of uncertainty about the belief analysis. It represents lack of evidence that can be due to lack of knowledge or expertise, or insufficient experimental analysis. The base rate distribution $\boldsymbol{a}_X$ is the prior probability distribution of X that reflects domain knowledge relevant to the specific analysis, most usually relevant statistical information.

A multinomial opinion can be represented as a point inside a regular simplex. In particular, a trinomial opinion can be represented inside a tetrahedron, as shown in Fig. 1. The beliefs are the distances to the sides and the uncertainty mass is the distance to the base of the tetrahedron. The base rate distribution is represented as a point on the base.

A subjective opinion in which $u_X = 0$, i.e. an opinion without any uncertainty, is called a *dogmatic opinion*. Dogmatic opinions correspond to probability distributions. A dogmatic opinion for which $\boldsymbol{b}_X(x) = 1$, for some $x \in \mathbb{X}$, is called

---

[1] This definition is for a *multinomial subjective opinion*. In general, we can define *hyper opinions*, where $\boldsymbol{b}_X : \mathcal{R}(\mathbb{X}) = 2^{\mathbb{X}} \setminus \{\mathbb{X}, \emptyset\}$, and operate with them through their multinomial projections (see [6]).

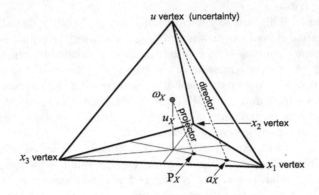

Fig. 1. Visualisation of a trinomial opinion

an *absolute opinion*. Absolute opinions correspond to observations. In contrast, an opinion for which $u_X = 1$ (and consequently $\boldsymbol{b}_X(x) = 0$, for every x) is called a *vacuous opinion*. For a given multinomial opinion ω_X we define its corresponding *projected probability distribution* $\mathbf{P}_X : \mathbb{X} \to [0,1]$ in the following way:

$$\mathbf{P}_X(x) = \boldsymbol{b}_X(x) + \boldsymbol{a}_X(x)\, u_X \ . \tag{3}$$

$\mathbf{P}_X(x)$ is an estimate for the probability of x which varies from the base rate value, in the case of complete ignorance ($u_X = 1$), to the belief in the case $u_X = 0$.

3   Joint Subjective Opinions

This section introduces a method of computing joint opinions. A *joint subjective opinion* on the variables $X_1, \ldots, X_n$, $n \geq 2$, is a tuple:

$$\omega_{X_1 \ldots X_n} = (\boldsymbol{b}_{X_1 \ldots X_n}, u_{X_1 \ldots X_n}, \boldsymbol{a}_{X_1 \ldots X_n})\,, \tag{4}$$

where $\boldsymbol{b}_{X_1 \ldots X_n} : \mathbb{X}_1 \times \ldots \times \mathbb{X}_n \to [0,1]$ and $u_{X_1 \ldots X_n} \in [0,1]$ satisfy the additivity condition in Eq. (2) and $\boldsymbol{a}_{X_1 \ldots X_n}$ is a joint probability distribution of the variables $X_1, \ldots, X_n$.

Our method assumes conditional opinions as part of the input. Given two (sets of) random variables X and Y, a *conditional opinion* on Y given that X takes the value x is a subjective opinion on Y defined as a tuple:

$$\omega_{Y|x} = (\boldsymbol{b}_{Y|x}, u_{Y|x}, \boldsymbol{a}_{Y|x})\,, \tag{5}$$

where the conditional belief mass distribution $\boldsymbol{b}_{Y|x} : \mathbb{Y} \to [0,1]$, the conditional uncertainty mass $u_{Y|x} \in [0,1]$ and the conditional base rate distribution $\boldsymbol{a}_{Y|x} : \mathbb{Y} \to [0,1]$ satisfy the additivity constraints of Eq. (2). We use the notation $\boldsymbol{\omega}_{Y|X}$ to denote a set of conditional opinions on Y, one for each value $x \in X$, i.e.:

$$\boldsymbol{\omega}_{Y|X} = \{\omega_{Y|x} \mid x \in \mathbb{X}\}\,. \tag{6}$$

3.1   Joint Subjective Opinion of Two Variables

Let X and Y be two random variables, and suppose we are given a subjective opinion ω_X and a set of conditional opinions $\omega_{Y|X}$. As we see later, this forms a two-node subjective Bayesian network. The computation of base rate distributions follows the same principles as the computation of traditional probability distributions in Bayesian networks.

We describe a method for determining a joint subjective opinion ω_{YX} on the Cartesian product of variables Y and X. This operation is denoted by the following expression:

$$\omega_{YX} = \omega_{Y|X} \cdot \omega_X \tag{7}$$

Determining Joint and Marginal Base Rates. In determining the base rates of the joint opinion ω_{YX}, we use the base rate distribution of X, a_X, and the projected probability distributions of the conditional opinions $\omega_{Y|x}$, $x \in \mathbb{X}$. This will also directly determine the marginal base rate distribution a_Y. For every $x \in \mathbb{X}$, and $y \in \mathbb{Y}$, the base rate is:

$$a_{YX}(y, x) = \mathbf{P}_{Y|x}(y) a_X(x) \tag{8}$$
$$= \left(b_{Y|x}(y) + a_Y(y) u_{Y|x} \right) a_X(x) \tag{9}$$

We assume that the marginal base rate distribution a_Y on the child variable Y can be obtained from the joint one by marginalizing out the variable X:

$$a_Y(y) = \sum_{x \in \mathbb{X}} a_{YX}(y, x) \tag{10}$$
$$= \sum_{x \in \mathbb{X}} \left(b_{Y|x}(y) + a_Y(y) u_{Y|x} \right) a_X(x) . \tag{11}$$

Solving the last equation for $a_Y(y)$, we obtain:

$$a_Y(y) = \frac{\sum_{x \in \mathbb{X}} b_{Y|x}(y) a_X(x)}{1 - \sum_{x \in \mathbb{X}} u_{Y|x} a_X(x)} , \tag{12}$$

Which along with Eq. (9) provides the joint and the marginal base rates.

3.2   Determining the Uncertainty and the Beliefs

First we compute the joint base rate distribution according to Eq. (9). Then we compute the projected probability distribution of the joint opinion and express it through Eq. (3):

$$\mathbf{P}_{YX}(y, x) = \mathbf{P}_{Y|x}(y) \mathbf{P}_X(x) \tag{13}$$
$$= b_{YX}(y, x) + a_{YX}(y, x) u_{YX} . \tag{14}$$

We impose the following requirement on the beliefs:

$$b_{YX}(y, x) \geq b_{Y|x}(y) b_X(x) . \tag{15}$$

This constraint comes from the interpretation of the belief masses as the minimum probability values (in which case the uncertainty mass will be a kind of 'non-assigned' probability mass that can distribute among the states of the domain in any possible way) and the fact that probability of the joint is a product of the probabilities $p_X(x)$ and $p_{Y|x}(y)$.

Now, having $\mathbf{P}_{YX}$ and a_{YX}, we need to find belief masses $b_{YX}(y, x)$ and uncertainty mass u_{YX} that satisfy Eq. (14) and the constraint in Eq. (15). We do this by looking for the maximum possible uncertainty mass value u_{YX} under the given constraints. From Eq. (14) we obtain:

$$u_{YX} = \frac{\mathbf{P}_{YX}(y, x) - b_{YX}(y, x)}{a_{YX}(y, x)}. \tag{16}$$

Applying Eqs. (3) and (13) to the last, we obtain:

$$u_{YX} = \frac{\left(b_{Y|x}(y) + a_Y(y)u_{Y|x}\right)b_X(x) + \left(b_{Y|x}(y) + a_Y(y)u_{Y|x}\right)a_X(x)u_X - b_{YX}(y, x)}{a_{YX}(y, x)}, \tag{17}$$

For every pair of values x and y, the maximum uncertainty mass on the right-hand side of Eq. (17) is achieved for the smallest allowable belief mass of $b_{YX}(y, x)$, which is $b_{Y|x}(y)b_X(x)$ by the Eq. (15). We denote that uncertainty mass by $u_{YX}(y, x)$:

$$u_{YX}(y, x) = \frac{a_Y(y)u_{Y|x}b_X(x) + b_{Y|x}(y)a_X(x)u_X + a_Y(y)a_X(x)u_{Y|x}u_X}{a_{YX}(y, x)}. \tag{18}$$

We take u_{YX} to be the minimum of the uncertainty masses in Eq. (18), i.e. $u_{YX} = \min_{y,x}[u_{YX}(y, x)]$, to assure that Eq. (15) always holds. In that way we obtain the following expression for the uncertainty of the joint opinion:

$$u_{YX} = \min_{y,x}\left[\frac{a_Y(y)u_{Y|x}b_X(x) + b_{Y|x}(y)a_X(x)u_X + a_Y(y)a_X(x)u_{Y|x}u_X}{a_{YX}(y, x)}\right]. \tag{19}$$

The joint belief mass distribution b_{YX} emerges from Eqs. (13) and (19):

$$b_{YX}(y, x) = \mathbf{P}_{YX}(y, x) - a_{YX}(y, x)\, u_{YX}. \tag{20}$$

The above described procedure generalizes for the case when Y has parents $X_1, \ldots, X_k$, where $k \geq 2$. Then the input arguments for the joint operation are: (1) a set of conditional opinions $\omega_{Y|X_1\ldots X_k}$ on Y, one for each combination of values of its parents; and (2) a joint opinion on the parents $\omega_{X_1\ldots X_k}$. In the next sections is shown that in a subjective Bayesian network (1) is given as input, and (2) is determined in a preceding step.

3.3   Joint Opinions for Multiple Variables

In this section we propose a way of extending the joint operation to more than two variables. Assume we have N random variables $X_1, \ldots, X_n$ where e.g. x_i represents a specific value of variable X_i. According to the chain rule of probability, we have the following expression for their joint probability distribution:

$$p(x_1, \ldots, x_n) = p(x_n|x_1, \ldots, x_{n-1}) \cdots p(x_2|x_1)p(x_1) , \qquad (21)$$

or written in a more compact way:

$$p(X_1, \ldots, X_n) = \prod_{k=1}^{n} p(X_k|X_{k-1} \ldots X_1) . \qquad (22)$$

We generalize Eq. (22) to subjective opinions:

$$\omega_{X_n \ldots X_1} = \prod_{k=1}^{n} \omega_{X_k|X_{k-1} \ldots X_1} , \qquad (23)$$

where the above product is performed applying the following recursive step:

$$\omega_{X_k \ldots X_1} = \omega_{X_k|X_{k-1} \ldots X_1} \cdot \omega_{X_{k-1} \ldots X_1} , \qquad (24)$$

for $k = 2, \ldots, n$, where the operation $\cdot$ is the joint of two subjective opinions. This procedure assumes that the conditional opinions $\omega_{X_k|X_{k-1} \ldots X_1}$ are available. This is the case when determining the joint opinions in a subjective Bayesian network, which we elaborate in the next section.

3.4   Subjective Bayesian Networks

A *Bayesian network* [11] with n variables is a directed acyclic graph (DAG) with random variables $V = \{X_1, \ldots, X_n\}$ as nodes, and a set of conditional probability distributions $\boldsymbol{p}(X_i|X_{\mathrm{Pa}(i)})$ associated with each node X_i containing one probability distribution $p(X_i|x_{\mathrm{Pa}(i)})$ of X_i for every value assignment $x_{\mathrm{Pa}(i)}$ of its parent nodes $X_{\mathrm{Pa}(i)}$.

If the Markov property holds for the given DAG and the joint distribution p of the variables $X_1, \ldots, X_n$ (every node is conditionally independent (I) of its non-descendant (ND) nodes given its parent (Pa) nodes in the graph, $\mathrm{I}\left(X_i, X_{\mathrm{ND}(i)}|X_{\mathrm{Pa}(i)}\right)$, then p is determined from the input information in the network as follows:

$$p(x_1, \ldots, x_n) = \prod_{i=1}^{n} p(x_i|x_{\mathrm{Pa}(i)}) , \qquad (25)$$

where $x_{\mathrm{Pa}(i)}$ is the value assignment of the parent variables of X_i that corresponds to the tuple $(x_1, \ldots, x_n)$.

A *subjective Bayesian network (SBN)* is a generalization of a classical Bayesian network where a set of conditional opinions $\omega_{X_i|X_{\mathrm{Pa}(i)}}$ substitutes the set of conditional probability distributions $p(X_i|X_{\mathrm{Pa}(i)})$. If we assume that the independencies embedded in the graph apply in the same way to the subjective opinions as to their corresponding probability distributions, then based on Eq. (25) we obtain the following equation:

$$\omega_{X_1,\ldots,X_n} = \prod_{i=1}^{n} \omega_{X_i|X_{\mathrm{Pa}(i)}} , \tag{26}$$

where the product denotes the joint operation. Assume $X_1,\ldots,X_n$ is an 'ancestral order' of the nodes in the graph meaning that there is no child that precedes its parent on the list, or more formally, if X_i is a parent of X_j, then $i < j$, where $i,j \in \{1,\ldots,n\}$. Then Eq. (26) is equivalent to Eq. (23) since for every i,

$$\omega_{X_i|X_{\mathrm{Pa}(i)}} = \omega_{X_i|X_{i-1}\ldots X_1} , \tag{27}$$

because $X_{i-1},\ldots,X_1$ are non-descendants of X_i and we assume that the Markov condition applies in the same way as for probabilities. The equality in Eq. (27) means that $\omega_{X_i|x_{i-1}\ldots x_1} = \omega_{X_i|(x_{\mathrm{Pa}(i)}, x_{\mathrm{NPa}(i)})}$ for every value assignment $x_{\mathrm{NPa}(i)}$ of the variables in the non-parent set $X_{\mathrm{NPa}(i)} = \{X_{i-1}\ldots X_1\} \setminus X_{\mathrm{Pa}(i)}$. Hence, we can apply subsequent joint operations to obtain the joint opinion on $X_1,\ldots,X_n$ using the information provided in the network. The same would apply for a subset of nodes $\{Y_1,\ldots,Y_k\} \subseteq V$ in the case they form an ancestral order that is *complete*, meaning that for every node Y_i all the ancestors of Y_i are contained on the list $Y_1,\ldots,Y_{i-1}$.

4 Opinion Marginalization

Let ω_{XY} be a joint multinomial opinion on the Cartesian product domain $\mathbb{X} \times \mathbb{Y}$. The joint opinion can be marginalized on one of the factor variables as either $\omega_{[X]}$ or $\omega_{[Y]}$. The special subscript notation $[\cdot]$ is used to indicate that the opinions are obtained through marginalization, and not, for example through operations like deduction described in Sect. 5. A marginal opinion represents the opinion on one of the joint factor variables without reference to the other factor variable, i.e. where the other variable has been *marginalized out*.

By applying probability marginalization and Eq. (3) we get:

$$\mathbf{P}_{[X]}(x) = \sum_{y \in \mathbb{Y}} \mathbf{P}_{XY}(x,y) , \tag{28}$$

$$b_{[X]}(x) + a_{[X]}(x)u_{[X]} = \sum_{y \in \mathbb{Y}} b_{XY}(x,y) + u_{XY} \sum_{y \in \mathbb{Y}} a_{XY}(x,y) . \tag{29}$$

Then we define:

$$a_{[X]}(x) = \sum_{y \in \mathbb{Y}} a_{XY}(x,y) , \tag{30}$$

$$b_{[X]}(x) = \sum_{y \in \mathbb{Y}} b_{XY}(x,y) , \tag{31}$$

$$u_{[X]} = u_{XY} . \tag{32}$$

The marginal opinion on X is the tuple $\omega_{[X]} = (\boldsymbol{b}_{[X]}, u_{[X]}, \boldsymbol{a}_{[X]})$. Similarly, $\omega_{[Y]}$ is obtained in the same way. The rationale for this definition of marginalization is that projected probabilities and base rates should behave like probability distributions, while the uncertainty of the marginal opinion should stay the same as in the joint opinion since the joint opinion is the only input in the operation (the amount/lack of information is the same). Then the beliefs inevitably marginalize in the same way as the probabilities.

4.1 Marginalization in SBNs

Opinion marginalization can be used in determining marginal opinions on one variable in a subjective Bayesian network. For simplicity we show this for a three-node network, but the generalization to an arbitrary network is straightforward. Assume a network e.g. expressed as $Z \to Y \to X$ where we want to determine ω_{XZ}.

To determine ω_{XZ} we marginalize the joint opinion ω_{XYZ} on the complete ancestral list X, Y, Z which is determined applying the joint operation. We obtain the following:

$$\mathbf{P}_{[XZ]}(x, z) = \sum_{y \in \mathbb{Y}} \mathbf{P}_{XYZ}(x, y, z) , \tag{33}$$

$$\boldsymbol{b}_{[XZ]}(x, z) = \sum_{y \in \mathbb{Y}} \boldsymbol{b}_{XYZ}(x, y, z) , \tag{34}$$

$$u_{[XZ]} = u_{XYZ} . \tag{35}$$

In the general case, if we want to determine the joint opinion on the variables $Y_1, \ldots, Y_k$ in a SBN, we first complete the sequence of nodes $Y_1, \ldots, Y_k$ to a complete ancestral order $X_1, \ldots, X_m$ by taking the ancestors of all the nodes in $\{Y_1, \ldots, Y_k\}$ and putting them in an ancestral order, and then determine the joint opinion on $X_1, \ldots, X_m$ according to the method described in Sect. 3.4. At the end, we marginalize it to obtain the opinion on the required set of variables.

5 Subjective Logic Deduction

This section briefly introduces subjective logic deduction as described in [3]. The method is based on the principle of maximizing the uncertainty upon constraints on the beliefs. In case the two variables are probabilistically independent, the joint operation reduces to the normal multiplication operation described in [8].

Given a set of conditional opinions $\boldsymbol{\omega}_{Y|X}$ and a subjective opinion ω_X, the goal of the deduction operation is to deduce a marginal opinion on Y, denoted $\omega_{Y\|X}$. Subjective logic deduction is denoted using the operator symbol '⊚' as:

$$\boldsymbol{\omega}_{Y\|X} = \boldsymbol{\omega}_{Y|X} ⊚ \omega_X . \tag{36}$$

The deduced opinion $\omega_{Y\|X}$ is a subjective opinion on Y. The special subscript notation $\cdot\|X$ is used to indicate that this marginal opinion has been computed through deduction.

First the projected probability distribution of $\omega_{Y\|X}$ is determined as follows:

$$\mathbf{P}_{Y\|X}(y) = \sum_{x \in \mathbb{X}} \mathbf{P}_{Y|x}(y)\mathbf{P}_X(x) . \tag{37}$$

The base rate distribution $\mathbf{a}_Y$ is determined by Eq. (12). It remains to determine the uncertainty and the beliefs of the deduced opinion.

For the belief masses of the deduced opinion $\omega_{Y\|X}$, we assume the following:

$$\mathbf{b}_{Y\|X}(y) \geq \min_{x \in \mathbb{X}} \left[\mathbf{b}_{Y|x}(y) \right] , \tag{38}$$

for every $y \in \mathbb{Y}$, which can be found as the *principle of plausible reasoning* in [12].

Let $\omega_{Y\|\widehat{X}}$ be the deduced opinion from the vacuous opinion $\widehat{\omega}_X$ with a base rate distribution $\mathbf{a}_X$ ($\widehat{u} = 1$). Then $u_{Y\|\widehat{X}}$ is determined as the maximum possible uncertainty mass value under the conditions imposed by Eqs. (37) and (38) applied to a vacuous opinion. The result is the following expression:

$$u_{Y\|\widehat{X}} = \min_{y \in \mathbb{Y}} \left[\frac{\sum_x \mathbf{P}_{Y|x}(y)\mathbf{a}_X(x) - \min_x \left[\mathbf{b}_{Y|x}(y) \right]}{\mathbf{a}_Y(y)} \right] . \tag{39}$$

The uncertainty of the opinion $\omega_{Y\|X}$ deduced from an arbitrary ω_X is then determined as the weighted average of the uncertainty mass $u_{Y\|\widehat{X}}$ and the uncertainty masses of the given conditional opinions:

$$u_{Y\|X} = u_X u_{Y\|\widehat{X}} + \sum_{x \in \mathbb{X}} \mathbf{b}_X(x) u_{Y|x} . \tag{40}$$

Equation (40) is the unique transformation that maps $\widehat{\omega}_X$ into $u_{Y\|\widehat{X}}$, and the corresponding absolute opinions on X into $u_{Y|x}$, for $x \in \mathbb{X}$. Once we have the uncertainty mass of the deduced opinion, the beliefs are easily derived as a consequence, applying Eq. (3) solved for $\mathbf{b}_{Y\|X}$:

$$\mathbf{b}_{Y\|X}(y) = \mathbf{P}_{Y\|X}(y) - \mathbf{a}_Y(y) u_{Y\|X} . \tag{41}$$

Deduction can also be generalized for the case when Y has parents $X_1 \dots X_k$, where $k \geq 2$, i.e. when the input arguments are a joint opinion $\omega_{X_1\dots X_k}$ and a set of conditional opinions $\omega_{Y|X_1\dots X_k}$ on Y, one for each combination of values of its parents.

6   Example: Marginalization vs. Deduction

The purpose of this section is to compare the direct and indirect method of computing the opinion on a child variable. The two methods are illustrated in Fig. 2. Note that $\omega_{Y\|X}$ and $\omega_{[Y]}$ in Fig. 2 are both subjective opinions on Y, where the subscripts indicate how they have been computed. The opinion $\omega_{Y\|X}$ has been computed through direct deduction, whereas $\omega_{[Y]}$ has been computed indirectly,

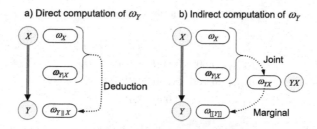

Fig. 2. The direct and indirect methods of computing ω_Y

by first computing the joint opinion on X and Y and then marginalizing out X. We compare the levels of uncertainty in these two opinions through the following example.

The scenario is a football match between Team 1 and Team 2, where it is generally predicted that Team 1 will win. A gambler who plans to bet on the match has received second-hand information about possible match-fixing whereby Team 1 has been paid to lose. The gambler has an opinion about the outcome of the match in case Team 1 has been paid to lose, and in the absence of match-fixing. The gambler also has an opinion about whether Team 1 actually has been paid to lose.

The variable X represents whether or not Team 1 has been paid to lose:

$$\mathbb{X} = \begin{cases} x_1 : & \text{Team 1 has been paid to lose,} \\ x_2 : & \text{No match-fixing.} \end{cases}$$

The gambler's opinion on it is given in Table 1. The provided base rate refers to a general rate of appearance of match-fixing.

Table 1. Opinion ω_X on the match-fixing

ω_X	x_1	x_2	u_X
b_X	0.90	0.00	0.10
a_X	0.10	0.90	

The variable Y represents the outcome of the game:

$$\mathbb{Y} = \begin{cases} y_1 : & \text{Team 1 wins the match,} \\ y_2 : & \text{Team 2 wins the match,} \\ y_3 : & \text{The match ends in a draw.} \end{cases}$$

Table 2 sets conditional opinions on Y given X, with marginal base rates from Eq. (12).

Table 2. Conditional opinions $\omega_{Y|X}$

	y_1	y_2	y_3	u_X
$b_{Y|x_1}$	0.000	0.800	0.100	0.100
$b_{Y|x_1}$	0.700	0.000	0.100	0.200
a_Y	0.778	0.099	0.123	

The first step in deriving the joint opinion ω_{YX} is to compute the joint base rate distribution using Eqs. (9) and (12) and the joint projected probability distribution according to Eq. (13), the results are provided in Table 3.

Table 3. (a) Joint base rate distribution, and (b) Joint projected probability distribution

a_{YX}	y_1	y_2	y_3
x_1	0.008	0.081	0.011
x_2	0.770	0.018	0.112

P_{YX}	y_1	y_2	y_3
x_1	0.063	0.656	0.091
x_2	0.162	0.004	0.024

The second step is to compute the joint uncertainty mass u_{YX} using Eq. (19) and then the joint belief mass distribution using Eq. (20). Table 4 provides the results.

Table 4. Joint belief mass distribution

b_{YX}	y_1	y_2	y_3	u_{YX}
x_1	0.062	0.646	0.090	0.120
x_2	0.070	0.002	0.010	

The derived marginal opinions on X and Y are shown in Table 5.

Table 5. (a) Marginal opinion on X, and (b) Marginal opinion on Y

	x_1	x_2	$u_{[X]}$
$b_{[X]}$	0.798	0.082	0.120
$a_{[X]}$	0.100	0.900	

	y_1	y_2	y_3	$u_{[Y]}$
$b_{[Y]}$	0.132	0.648	0.100	0.120
$a_{[Y]}$	0.778	0.099	0.123	

Next we determine the deduced opinion $\omega_{Y\|X}$ based on the given ω_X and the set of conditional opinions $\omega_{Y|X}$. First, the marginal base rate distribution a_Y is determined as explained in Sect. 3.1. The second step is to use Eq. (39) to compute the uncertainty $u_{Y\|\hat{X}} = 0.29$. The third step is to apply Eqs. (40) and (41) to compute the deduced opinion about which team will win the match. The results are in Table 6.

Table 6. Deduced opinion on Y

	y_1	y_2	y_3	$u_{Y\|X}$
$b_{Y\|X}$	0.133	0.648	0.100	0.119
a_Y	0.778	0.099	0.123	
$\mathbf{P}_{Y\|X}$	0.225	0.660	0.115	

Based on the opinion about match-fixing, as well as on the conditional opinions, it appears that Team 1 has a relatively slim chance of winning. Despite the high marginal base rate of winning given by $a_Y(y_1) = 0.778$, when the evidence of match fixing is taken into account the projected probability of Team 1 winning the match is only $\mathbf{P}_{Y\|X}(y_1) = 0.225$. The probability of Team 2 winning is $\mathbf{P}_{Y\|X}(y_2) = 0.660$. The uncertainty mass in the marginal opinion $u_{[Y]} = 0.120$ is not exactly the same as the deduced uncertainty mass $u_{Y\|X} = 0.119$.

It can also be seen that the marginal opinion $\omega_{[X]}$ on X in Table 5 is close, but not exactly the same as the original opinion ω_X in Table 1.

7 Discussion and Conclusions

We have shown that the notion of joint probability distributions can be extended to joint subjective opinions, providing a method to determine the joint opinion of two variables and a way to extend it to multiple variables.

In a subjective Bayesian network with two variables, the obtained marginalized opinions $\omega_{[X]}$ and $\omega_{[Y]}$ can be compared against the input opinion ω_X and the deduced opinion $\omega_{Y\|X}$, respectively. Since the base rate distributions and the projected probability distributions are determined in the same way in both the cases, and the rest relies on the uncertainty mass of the opinion, that is exactly what we are observing in this comparison. The difference between the two uncertainty values is due to the approximate nature of uncertainty mass in the computation of deduction and joint opinions, which makes the direct method (the deduction) result in an opinion with slightly less uncertainty mass than the indirect method (marginalization of the joint opinion). However, the example shows that this difference is relatively small, so the two methods produce very similar results.

A method for computing joint opinions involving an unnecessary indirect step of applying the subjective Bayes' theorem is discussed in [6]. Due to limited space, and because it has limited interest, that method is not described here. Suffice to say that it produced relatively poor approximations of the uncertainty level of the joint opinion in a similar example to the one described above. In contrast, the method presented here computes the joint opinion directly based on the principle of maximizing the uncertainty, and in that sense is more compatible and comparable with both multiplication and deduction that are derived based on the same principle. The method presented here produces very good approximations of the uncertainty level in the specific example described. A general analytical investigation of the quality of the approximation of the uncertainty level is difficult because the expressions have many degrees of freedom. A partial analysis of the approximation quality of the product operator was done in [5], which showed very high quality, and we assume the same to be the case for the joint operator presented here.

There exists a bijective mapping between subjective opinions and Dirichlet probability density functions (PDFs) [7]. This correspondence is widely exploited for reducing the inference with subjective opinions to approximate inference with the corresponding Dirichlet PDFs [9,10]. In that sense, the method for joint opinions described here is also a method for approximating joint Dirichlet PDFs. The joining of two Dirichlet PDFs does not produce a Dirichlet PDF in general, but a hypergeometric PDF which has different higher-order moments than the Dirichlet PDF. However, to derive the analytically correct expressions would be intractable in the general case. The joint operator described here approximates the result to a Dirichlet PDF. The advantage is its simplicity and low computational complexity. The simplicity of the method is therefore a trade-off with the approximate results it produces.

References

1. Cozman, F.G.: Credal networks. Artif. Intell. **120**(2), 199–233 (2000)
2. Ivanovska, M., Jøsang, A., Sambo, F.: Subjective networks: prospectives and challenges. In: Third International Workshop on Graph Structures for Knowledge Representation and Reasoning, GKR@IJCAI 2015, Buenos Aires, July 2015
3. Ivanovska, M., Jøsang, A., Sambo, F.: Bayesian deduction with subjective opinions. In: Delgrande, J., Wolter, F., Baral, C. (eds.) 15th International Conference on Principles of Knowledge Representation and Reasoning, KR 2016. AAAI Press, Palo Alto (2016)
4. Jøsang, A.: Conditional reasoning with subjective logic. J. Multiple Valued Logic Soft Comput. **15**(1), 5–38 (2008)
5. Jøsang, A., McAnally, D.: Multiplication and comultiplication of beliefs. Int. J. Approximate Reasoning **38**(1), 19–51 (2004)
6. Jøsang, A.: Subjective Logic: A Formalism for Reasoning Under Uncertainty. Springer, Heidelberg (2016)
7. Jøsang, A., Hankin, R.: Interpretation and fusion of hyper-opinions in subjective logic. In: Proceedings of the 15th International Conference on Information Fusion, FUSION 2012, Singapore. IEEE, Los Alamitos, July 2012
8. Jøsang, A., O'Hara, S.: Multiplication of multinomial subjective opinions. In: International Conference on Information Processing and Management of Uncertainty, IPMU 2010, Dortmund (2010)

9. Kaplan, L., Ivanovska, M.: Efficient subjective Bayesian network belief propagation for trees. In: Proceedings of the 19th International Conference on Information Fusion, FUSION 2016. IEEE, Los Alamitos (2016)
10. Kaplan, L.M., Sensoy, M., Tang, Y., Chakraborty, S., Bisdikian, C., de Mel, G.: Reasoning under uncertainty: variations of subjective logic deduction. In: Proceedings of the 16th International Conference on Information Fusion, FUSION 2013, Istanbul, Turkey, 9–12 July 2013, pp. 1910–1917 (2013)
11. Pearl, J.: Probabilistic Reasoning in Intelligent Systems. Morgan Kaufmann, Burlington (1988)
12. Pearl, Judea: Reasoning with belief functions: an analysis of compatibility. Int. J. Approximate Reasoning 4, 363–389 (1990)
13. Walley, P.: Statistical Reasoning with Imprecise Probabilities. Chapman and Hall, London (1991)

Author Index

Printed in the United States
By Bookmasters